AGRICULTURAL ENGINEERING DEPARTMENT
UNIVERSITY OF MINNESOTA
INSTITUTE OF AGRICULTURE
ST. PAUL, MINNESOTA
55101

0100 - 2212 - 02

REC'D 6/1/71

PILOT PLANTS, MODELS, AND SCALE-UP METHODS IN CHEMICAL ENGINEERING

BUILDING THE LITERATURE OF A PROFESSION

Fifteen prominent chemical engineers first met in New York more than 30 years ago to plan a continuing literature for their rapidly growing profession. From industry came such pioneer practitioners as Leo H. Baekeland, Arthur D. Little, Charles L. Reese, John V. N. Dorr, M. C. Whitaker, and R. S. McBride. From the universities came such eminent educators as William H. Walker, Alfred H. White, D. D. Jackson, J. H. James, Warren K. Lewis, and Harry A. Curtis. H. C. Parmelee, then editor of *Chemical & Metallurgical Engineering*, served as chairman and was joined subsequently by S. D. Kirkpatrick as consulting editor.

After several meetings, this first Editorial Advisory Board submitted its report to the McGraw-Hill Book Company in September, 1925. In it were detailed specifications for a correlated series of more than a dozen texts and reference books which have since become the McGraw-Hill Series in Chemical Engineering.

Since its origin the Editorial Advisory Board has been benefited by the guidance and continuing interest of such other distinguished chemical engineers as Manson Benedict, John R. Callaham, Arthur W. Hixson, H. Fraser Johnstone, Webster N. Jones, Paul D. V. Manning, Albert E. Marshall, Charles M. A. Stine, Edward R. Weidlein, and Walter G. Whitman. No small measure of credit is due not only to the pioneering members of the original board but also to those engineering educators and industrialists who have succeeded them in the task of building a permanent literature for the chemical engineering profession.

Pilot Plants, Models, and Scale-up Methods in Chemical Engineering

ROBERT EDGEWORTH JOHNSTONE

Assistant Director of Ordnance Factories
Ministry of Supply, London

MEREDITH WOOLDRIDGE THRING

Professor of Fuel Technology and Chemical Engineering
University of Sheffield

McGRAW-HILL BOOK COMPANY

New York   Toronto   London

1957

PILOT PLANTS, MODELS, AND SCALE-UP METHODS
IN CHEMICAL ENGINEERING

Library of Congress Catalog Card Number 56-11049

7 8 9 10 11 12 – MAMM – 7 6 5 4 3 2 1 0

32693

This book is dedicated
to the memory of
GEORGE E. DAVIS
whose classical two-volume work
"A Handbook of Chemical Engineering"
was published by Davis Bros. of Manchester, England,
in the year 1901

Vitruvius says that small models are of no avail for ascertaining the effects of large ones; and I here propose to prove that this conclusion is a false one.

Leonardo da Vinci, *Notebooks* (about A.D. 1500)

PREFACE

This book is an attempt to systematize and present in a usable form such quantitative methods as are available for predicting the performance of large-scale process plant from small-scale experiments. These methods can be of value in two distinct fields: first, the transference of new processes from pilot plant to large-scale operation, and secondly, the study of the behavior of existing full-sized plant units by means of suitable models. A good deal has been written on both of these topics, especially the first, but nearly always from a descriptive and qualitative point of view. It was felt that the time had come for an analytical and quantitative approach to the whole subject.

The book contains a fair amount of mathematics, but with the possible exception of the last chapter none of it is very advanced. It is hoped that enough descriptive matter and numerical examples have been included to render the mathematical derivations easily intelligible. Moreover, most of the results have been reduced to the form of simple *scale equations* which can be applied directly to practical cases. These equations express ratios of performance figures in model and prototype under specified operating conditions.

Model theory as applied to process plant is a relatively new field, and so it has been necessary to set up a terminology which is in some respects novel. This terminology is summarized in Appendix 1, and the reader is recommended to glance through it before attempting any serious study of the book itself. In particular, the specialized use of the term *homologous* should be noted.

The authors are grateful to all those who have helped in the production of this book, and especially to Professor K. G. Denbigh of the University of Edinburgh and Dr. H. R. C. Pratt of the Atomic Energy Research Establishment, Harwell, England, who have read and commented upon Chaps. 6 and 13, respectively. They are not, of course, to be held responsible for any of the statements contained therein. Thanks are also due to Mrs. E. E. Arundel, who has given spare-time secretarial services for several years.

In 1938 one of the authors sent a questionnaire to some of the lead-

ix

ing manufacturers of process plant in Great Britain and the United States, inquiring whether and how far they made use of models. From the replies received, it appeared that models were already being employed to some extent in plant design, although correlation with large-scale performance was almost entirely empirical. Shortly afterward war broke out, and the work on model theory had to be set aside, but perhaps even now it is not too late to thank publicly those firms which completed and returned the questionnaire.

The authors' principal debt, however, is to the various industries with which they have been personally associated for a combined total of more than 50 years and which include coal, explosives, heavy chemicals, iron and steel, petroleum, and plant construction. The experience gained in these industries both suggested the need for the present work and governed the selection and presentation of material. It is the authors' aim and hope to have contributed something useful to the common pool of chemical-engineering knowledge from which they have drawn so fruitfully in the past.

ROBERT E. JOHNSTONE
MEREDITH W. THRING

CONTENTS

CHAPTER 1

INTRODUCTION

The ultimate purpose of all pilot-plant and model experiments is crystallized in a phrase by L. H. Baekeland that has become famous: "Commit your blunders on a small scale and make your profits on a large scale" (228).* George E. Davis, author of the world's first handbook of chemical engineering, emphasized the value of experiments on a scale intermediate between that of the laboratory and full-scale production. "A small experiment made upon a few grammes of material in the laboratory will not be much use in guiding to the erection of a large scale works, but there is no doubt that an experiment based upon a few kilogrammes will give nearly all the data required . . . " (134).

Today there appears to be a trend toward the omission, under certain circumstances, of the pilot-plant stage in the development of new processes. An early instance was the development of the hot-acid alkylation process described in 1937 by McAllister (22); the most spectacular example was the wartime production of the atomic bomb. A full-scale plant that is designed entirely without previous pilot-scale experimentation is still the exception; yet if the present trend were to continue unchecked, a book on pilot plants would eventually become superfluous. It is therefore appropriate to begin by reviewing the considerations which have led the authors to offer such a work at this time.

There is little doubt that it would be technically possible to transfer any new process whatever from the laboratory directly to large-scale production provided that unlimited money were available, so that the designer could allow huge factors of safety and the operator could if necessary meet the cost of overcoming massive "teething" troubles. On the other hand, there is also little doubt that performance data obtained from a correctly designed and operated small-scale plant are always more accurate and reliable than data scaled up from the labora-

* Numbers in parentheses refer to the Classified References at the back of the book.

1

tory bench or derived by calculation from generalized correlations. Smaller factors of safety can be allowed in the design of the production plant, yields and efficiencies will often be better, and there are likely to be fewer initial troubles on the large scale than if the plant had been wholly designed on paper. For a commercial process that is required to earn profits, the decision on whether to omit the pilot-plant stage or not must rest upon an economic balance in which savings in time and development costs are set against higher efficiency and lower capital and start-up costs. The point of balance will vary with the growth of theoretical knowledge on the one hand and improvements in experimental methods on the other.

From the beginnings of chemical engineering up to the present time, the development of theoretical and semiempirical methods of calculation has outrun improvements in experimental technique. Design calculations have advanced from the simple stoichiometry of Davis's time to elaborate mathematical procedures derived from theoretical and practical studies of the unit operations. Pilot-plant experiments, on the other hand, are still likely to be conducted on trial-and-error principles very much as they might have been conducted fifty years ago. Many useful papers and articles have been written about pilot-plant design (230, 231, 237, 238), pilot-plant operation (232, 234, 239), and process development generally (229, 233, 236, 240 to 243, 245), but the treatment has nearly always been descriptive rather than quantitative. In these circumstances, it is natural that the trend up to now should have been toward more calculation and less experiment.

There are signs that this phase may be coming to an end and that in the years ahead more attention will be given to the elaboration and refinement of experimental techniques able to furnish quickly and at low cost design data of a higher order of accuracy than that obtainable from generalized correlations. The first important move in this direction was the application to chemical engineering of statistical methods for the design of experiments. Brownlee's monograph of 1946 (129) was one of the first systematic adaptations of these methods to the special problems of the plant chemist and chemical engineer. Other and more detailed works have since appeared (132, 133). Pilot-plant experiments are particularly costly and time-consuming, and a suitable experimental design and statistical analysis of the results can greatly reduce the number of test runs that would otherwise be required to produce a given amount of information. Today, no development engineer can afford to neglect the statistical approach. A simple design that could be adapted to many chemical-engineering investigations is described by Gore (137).

A second development in experimental technique that is beginning to attract attention is the application of model theory to the scaling up (or down) of chemical plant and processes. It is a well-known fact that small-scale experiments are not always a reliable guide to large-scale results. Sometimes a process which is satisfactory in the pilot plant gives trouble on the large scale, and occasionally the reverse is found. The aim of model theory is to predict these scale effects and determine the conditions (if any) under which the performance of a model gives a reliable forecast of prototype performance. Quantitative methods of correlating the behavior of model and full-scale systems have been successful in other fields of engineering, notably the design of ships and aircraft, where processes occur that are too complex for mathematical analysis. Applied to chemical plant, such methods point to the possibility of larger scale-up factors, smaller and cheaper pilot plants, and more dependable results.

It seems possible that these two experimental techniques, statistical analysis and model theory, may together serve to arrest the drift away from pilot-plant trials and toward paper work, or even reverse it, so that small and easily made models will be used for the accurate determination of parameters that are now approximated by calculation. The economic point of balance would then begin to move in the opposite direction. Another factor militating against the complete disappearance of the pilot plant is that the provision of design and operation data is not the only function of pilot-plant experiments. Very often it is necessary to establish trickle production of a new chemical in order to explore its uses and market. In such cases, a pilot plant would be set up even though all the technical data needed for a full-scale plant were already available. And, being needed in any case, the pilot plant could advantageously be used to furnish additional and more reliable design data by the application of model theory.

It is significant that, while in recent years there has been some tendency to curtail or even to omit the pilot-plant stage in developing new processes, an opposite trend is seen in relation to established processes, namely, the increasing use of models to predict the effects of changes of design or operating conditions in existing large-scale plant. Model techniques have been particularly valuable in relation to the design and performance of furnaces, and they are also largely used in the sphere of mixing equipment. A recent development is the analogue model, which may, for example, simulate hydrodynamic processes electrically, or vice versa (see Chap. 19). In its most advanced form, it becomes an analogue computer, and at this point the distinction between calculation and experiment vanishes.

Summing up, one may conclude that, first, there will always be circumstances in which pilot-plant or model experiments are economically justified. Second, improvements in experimental technique may well reverse the current trend and lead to more pilot plants rather than fewer. Third, the use of models is already on the increase.

The applicability of model theory to chemical-engineering problems was pointed out by one of the authors in 1930 (250a), but the first specific application appears to have been Damköhler's study of the continuous chemical reactor published in 1936 (7). Since that time, various papers on the subject have appeared, at first occasionally and lately with increasing frequency. Even so, the literature explicitly dealing with chemical-engineering model theory is still scanty. Much more important is the model theory which is implicit in the many dimensionless correlations already used in chemical engineering. In a sense, we have all been model theorists without knowing it. Every time a research worker formulates his experimental results in terms of dimensionless groups he tacitly assumes a scale-up rule. In order to scale up a particular system with which one has already experimented, there is no need to go to the trouble of determining or guessing the physical properties of the substances present and evaluating the dimensionless groups. It is necessary only to extract from the dimensionless correlation the appropriate scale-up rule and use that directly.

In view of the trends and needs described above, the authors felt that it would be a useful task to collect, develop, and systematize all aspects of current model theory, either explicit or implicit, which could be of use in chemical engineering. The simplicity of the scale relations which emerge will appeal both to students and to practical development engineers. It is hoped that teachers of chemical engineering may find in model theory a useful supplement to the analytical treatment of the subject and that its wider application in industry may lead the way to smaller and cheaper pilot plants and models giving more and better information in a shorter time.

CHAPTER 2

PILOT PLANTS AND MODELS

Small-scale plant is employed in chemical engineering for two main purposes. The first is as forerunner to a full-sized production plant that is not yet built. In this case, the small-scale equipment is called a *pilot plant*,* and its principal function is usually to provide design data for the ultimate large one, although it may also be required to produce small quantities of a new product for trial. The second purpose is to study the behavior of an existing plant of which the small unit is a reproduction. In this case, the small-scale equipment is what is ordinarily called a *model*, and its chief function is to exhibit the effects of change in shape or operating conditions more quickly and economically than would be possible by experiments on the full-sized prototype. The functions of a pilot plant belong to the sphere of process development, those of a model of an existing plant to the sphere of process study. For the present purpose, it is not important whether the small-scale unit is the forerunner from which a full-sized counterpart will ultimately be scaled up, as in the case of a pilot plant, or whether the small unit is itself a scaled-down model of an existing piece of plant. The concern is with the relative performance of corresponding small- and large-scale units, irrespective of which comes first in time. In the following chapters, the term model will be given a purely geometrical interpretation with respect to actual or possible large-scale prototypes or replicas. It is generally desirable that a pilot plant should be designed with some particular type and size of large-scale unit in mind, and in this sense the pilot plant may be regarded as a model of an imaginary prototype.

* Some writers on process development distinguish between *pilot plants, semi-technical plants, semicommercial plants,* etc., according to the scale and purpose of operation. In the terminology of this book these are all pilot plants. Neither the scale of operation nor the purpose for which it is carried on affect the model laws that apply to a particular process.

5

Process Development

For purposes of plant design, a pilot plant is properly employed to gain such information as cannot be obtained by any cheaper or quicker method. In the development of a new chemical process, it is seldom necessary to reproduce the entire process on the pilot-plant scale in order to obtain design data. Only those plant items need to be "piloted" which cannot be designed sufficiently closely from past experience or from known principles on the basis of laboratory data. How these items are arrived at may be illustrated by a brief discussion of the methods of process development.

Nowadays, the chemical engineer is usually asked to make a first evaluation of a new process while it is still in the laboratory stage. This he does by carrying out a *preliminary engineering study*. Flow diagrams are prepared showing material quantities and any information on heat or power requirements that is available. On the basis of laboratory results, the complete process is broken down into a series of unit operations or processes, and a suitable type of equipment for each step is tentatively chosen. At this stage, a rough estimate of capital cost may be attempted (the estimation of costs lies outside the scope of the present work, but it may be mentioned that in many cases the total capital investment in a new plant amounts to roughly four times the cost of the major plant items as estimated from the preliminary engineering study).

Having tentatively selected the type of equipment to be used, the chemical engineer is able to write down all the quantitative data that will be needed for the design of each component. Such data will include material and heat balances, chemical, physical, and thermodynamic properties of raw materials, intermediate compounds and final products, reaction rates, heat- and mass-transfer coefficients and other rate parameters, power requirements, corrosion rates, etc. The design data required for each piece of equipment can then be classified under six heads:

1. Data available from past experience.

2. Data given in the laboratory reports or which can be derived from the laboratory results.

3. Data available in the literature.

4. Data which can be approximated sufficiently closely for design purposes by means of thermodynamic relations, the theorem of corresponding states, or some of the many empirical or semiempirical correlations that have appeared in recent years.

5. Data which could be obtained by further research in the laboratory.

6. Other practical data felt to be necessary for design purposes. Information falling under this head can be determined only in a pilot plant.

Occasionally the preliminary engineering study may suggest that no pilot-scale experiments at all are needed. The new process may be entirely made up of standard unit operations which are well understood and from which plant can be confidently designed from laboratory data alone. Usually such a procedure, when it is possible, necessitates relatively large margins of safety and correspondingly costly plant, and there is also the risk of unsatisfactory performance owing to some unexpected scale effect (see Chap. 1).

As a rule, therefore, one or two "critical" operations of the new process are selected for study on a pilot-plant scale. In order to yield the maximum amount of useful information, the experimental equipment for the critical operations requires to be designed on correct principles and operated under appropriate conditions. If it should be necessary to carry out other steps in the process in order to provide material for the critical operations or deal with the products therefrom, the plant for such auxiliary operations need not be specially designed for the purpose but may consist of any general-utility or improvised equipment of suitable size that happens to be available. This equipment is not required to furnish design data for its large-scale counterpart, and so its efficiency and performance are relatively unimportant, although useful supplementary information can often be obtained from it. In testing a continuous process on the pilot scale, it is sometimes advantageous to conduct the preliminary steps on the batch system, accumulate enough intermediate product to permit a continuous run on that section of the process in which operations are critical, and afterward submit the output of the continuous run to further treatment by batch processes in order to obtain the final product. This procedure may enable the cost of the pilot plant to be substantially reduced, and there is the additional advantage that the pilot-plant operator is free to devote his whole attention to the critical operations while they are in progress. Somewhat similar considerations apply to models of existing equipment. It is not always necessary to set up a complete miniature reproduction of the piece of plant to be studied. Sufficient information can often be obtained from a model representing only a part of the complete unit (for example the "slice" models of furnaces mentioned in Chap. 16).

A pilot plant is unlikely to yield the maximum possible amount of information unless the "critical" components at least are designed and operated in accordance with model theory. The first step is to derive the similarity criteria which govern the operations or processes to be studied on the small scale. These may be obtained either by dimensional analysis or from the fundamental differential equations of the process, as described in Chaps. 4 and 5. A study of the similarity criteria will reveal the conditions under which the model should be tested in order that the results may simulate those obtained under given conditions on the large scale. It will also show whether there are likely to be appreciable scale effects, in which case it will be necessary to apply corrections to the pilot-plant results before they are used for full-scale design. Finally, such a study will reveal those awkward cases in which two or more similarity criteria are incompatible and where, consequently, it is not possible to simulate large-scale results in a small apparatus with any certainty. Even in these cases, experiments with a small-scale unit over a sufficiently wide range of conditions can give valuable information for the design of the full-sized plant.

The safest method of evaluating scale effects and allowing for incompatibility of similarity criteria is to scale a process up by easy stages through two or three pilot plants of increasing size. The effects, if any, of change of scale on rate parameters and yields can then be observed directly and extrapolated to the full scale. A graduated succession of pilot plants is part of the traditional technique of process development, a technique that is both costly and slow. The principal aim of model theory is to allow of larger steps in development and reduce their number and duration, and it is to be hoped that in the future the need for more than one pilot stage between the laboratory and the full scale will become increasingly rare.

It is often difficult to decide how much money ought to be spent on developing a new process before it is exploited. One can continue experimenting and improving the process forever, but after a time any possible future gains are more than offset by the current loss of revenue through delay in exploitation. Yates (144) has given a statistical analysis of situations of this kind. Making certain assumptions, he concludes that the optimum expenditure on experimentation is reached when the direct cost (excluding overhead) is equal to the expectation of loss due to imperfections remaining in the process, i.e., equal to the difference between the total anticipated profits from the plant that could actually be built and those from a theoretically perfect plant.*

---

* Yates's examples were taken from agriculture, but his principle is of general application and worthy of study by those interested in the economics of industrial research. His main assumption is that in the vicinity of the optimum the economic loss varies as the square of the deviation from the optimum.

Process Study

By process study is meant the examination and improvement of existing full-scale processes or plant. Here the principal object of using models is to determine the effect of modifications in design or major changes in operating conditions without incurring the expense or risk of making these changes on the large scale. For example, to establish the effect of altering the shape or position of a baffle wall in a furnace by means of a full-scale trial, it would be necessary to take the furnace out of service, allow it to cool down, break out the old baffle wall, build the new one, and heat up sufficiently slowly to avoid damaging the brickwork. Altogether the furnace might be out of use for a month or more, and again perhaps for an equal period of restoration to its original state if the change should have proved deleterious. The required information could probably be obtained by means of a model in a few days and without interrupting production.

Similarly, a major change in operating conditions which might conceivably damage a production plant, perhaps coke up a tube still or sinter a catalyst mass, can be tested first on a small scale provided that a model apparatus is available which can be relied upon to reproduce the behavior of the prototype. Model theory indicates the necessary conditions for such reproducibility. A working model of a production plant can furnish valuable advance data on the effect of proposed changes in operation or hookup. For this reason, it is often wise to keep a pilot plant in commission long after its original process-development function has been accomplished. As the detailed design and construction of the full-scale plant proceed, the pilot plant is then available to solve last-minute problems. In the case of a difficult or dangerous process, it can provide a training ground for production-plant operatives. Finally, when the full-scale plant is built and in production, the pilot plant becomes a working model in which the characteristics of the process can be further studied at relatively little cost. More than once, in the authors' experience, a company has had reason to regret the too hasty dismantling of a pilot plant.

Errors Avoided

In order to evaluate the gain in accuracy and dependability of data to be expected from pilot-plant and model studies, it is necessary to examine the possible sources of error in the alternative procedure of calculating plant dimensions from laboratory data and theoretical or empirical design equations. The fundamental type of design equation is the *rate equation*, which gives the rate at which a physical or chemical

process takes place and hence the size of equipment needed to effect
the required degree of change in a given throughput.   The quantities
which enter into a typical rate equation may be divided into four
groups:
1. Dimensional and time factors
2. Operating variables
3. Physical or chemical properties
4. Numerical factor

The *dimensional factor* indicates the manner in which the total
quantity of change per hour depends upon the dimensions of the
apparatus—whether, for example, it varies with the length, surface, or
volume of the vessel.   The *time factor* indicates the manner in which
the rate varies with time.   For processes proceeding at a constant rate,
both dimensional and time factors are commonly incorporated in a
rate coefficient such as the heat-transfer coefficient, mass-transfer
coefficient, or chemical-reaction velocity constant.   For rates which
vary with time, the time factor enters as a differential.

Operating variables include temperatures, pressures, concentrations,
rates of flow, and residence times.   In general, they constitute the
independent variables which are under the plant operator's control.

Physical or chemical properties represent the usual form in which
laboratory data are fed into the design calculation.   They may be
determined in the laboratory for this purpose, taken from the literature
or reference handbooks, or approximated by means of some theoretical
or empirical rule.

The *numerical factor* is a dimensionless factor which is needed to
balance the rate equation when the latter is expressed in dimensionless
form.   An example is the factor of 0.023 in the Dittus-Boelter equation for heat transfer inside pipes.   This numerical factor is actually a
shape factor which varies with the geometrical shape of the apparatus
but is independent of its size.   For example, for external flow across
tube banks the numerical factor for heat transfer becomes 0.33.   Both
factors, 0.023 and 0.33, are independent of tube diameter, but the
latter figure is affected by the way the tubes are arranged, whether
staggered or in line.

Each of the above classes of quantity represents a possible source of
error when introduced into the rate equation without experiment.
For example, even the dimensional factor is uncertain in the case of
transfer processes involving a fluid interface.   The common practice
is to work with a volume coefficient, but this in effect assumes a certain
specific surface per unit volume which may not be the same in the
proposed plant as it was in the experiments from which the coefficient

was derived. Operating variables are perhaps the least uncertain quantities in the rate equation, though values taken from the literature may have been measured by methods differing from those which will be used and giving different results. Physical and chemical properties derived from the literature are by no means free from error even when found in reputable handbooks. Some handbook data are still based on early experiments, and it is surprising how serious errors can persist through the years. Properties approximated by means of empirical rules are still more uncertain. Finally, most of the dimensionless rate equations for physical processes are strictly valid only for apparatus of a particular shape. Thus, the factor of 0.023 in the Dittus-Boelter equation applies only to straight pipe, and even then it varies with the ratio of length to diameter. In short pipes, the effect of this ratio is appreciable.

In a correctly planned pilot-plant or model experiment, most of the above sources of error are avoided.* Operating variables may be measured by the same methods as those which will be employed on the large scale. If the same process materials are treated under the same conditions in the pilot plant as on the large scale, it is not necessary to assume values for physical and chemical properties, since in scaling up the plant performance those quantities cancel out. (Model experiments in which different materials are processed on the small scale, such as a water model of a furnace, are still dependent on an accurate knowledge of physical properties.) Finally, where the appropriate geometrical relations exist between small- and large-scale apparatus, the numerical shape factor cancels out. As a result, under correctly chosen conditions the scale-up relation between a model and its prototype can generally be reduced to a simple expression containing only ratios of linear dimensions and operating variables.

* Sources of error are further discussed in Chap. 8, second paragraph onward.

CHAPTER 3

THE PRINCIPLE OF SIMILARITY

The principle of similarity is concerned with the relations between physical systems of different sizes, and it is thus fundamental to the scaling up or down of physical and chemical processes. The principle was first enunciated by Newton for systems composed of solid particles in motion (253). The earliest practical applications were to fluid systems, and it is in this field that the principle has proved itself to be most useful. By the seventies of the last century, W. Froude was applying similarity criteria to the prediction of wave drag on ships' hulls from experiments with models, a method that is still used. Soon afterward, Osborne Reynolds employed models to investigate the erosion of estuaries and river beds. During 1914 and 1915, there was a series of papers and letters by Tolman (259), Buckingham (247, 248), Rayleigh (254, 255), and Riabouchinsky (256) in which the wider applications and fundamental significance of the principle of similarity were described and debated. Since that time, an important development has been the prediction of aircraft performance from experiments with models. The theory of models as applied to mechanical- and civil-engineering problems is very thoroughly dealt with in a recent text by Langhaar (251).

In the province of chemical engineering, the chief practical use of similarity up to the present has been to correlate the performance of geometrically similar paddle, propeller, and turbine mixers. Early systematic studies in this field were made by Hixson and his coworkers (191 to 197), and the work has been continued by others, notably Rushton (206 to 208). Applications of the principle of similarity to chemically reacting systems have been dealt with by Damköhler (7, 8), Edgeworth Johnstone (18), Bosworth (1 to 4), and Thring (64, 258). The last author used the principle to solve various practical problems in furnace design.

The principle of similarity is usually coupled, and not infrequently confused, with the method of dimensional analysis. Although his-

12

torically the two have been linked together, yet logically they are quite distinct. The principle of similarity is a general principle of nature: dimensional analysis is only one of the techniques by which the principle may be applied to specific cases, the other technique being to start from the generalized equations of motion of the system. In this and the following chapters, similarity and dimensional analysis are discussed separately, while equations of motion have a further chapter to themselves. The object of the present chapter is to show the physical significance of similar states. The requisite conditions for similarity under different regimes are dealt with in Chap. 7.

Material objects and physical systems in general are characterized by three qualities: size, shape, and composition. All three are independently variable, so that two objects may differ in size while having the same shape and chemical composition, or they can be alike in shape only but of different sizes and composed of different materials. The principle of similarity is more particularly concerned with the general concept of *shape* as applied to complex systems and with the implications of the fact that shape is independent of size and composition. In more precise terms, this principle states that *the spatial and temporal configuration of a physical system is determined by ratios of magnitudes within the system itself and does not depend upon the size or nature of the units in which these magnitudes are measured.*

The chemical engineer is concerned with complex systems composed of solid bodies and fluids in which transfer of matter and energy may take place as well as chemical change. The concept of "shape" applied to these systems involves not only the geometrical proportions of their solid members and surfaces but also such factors as fluid-flow patterns, temperature gradients, time-concentration profiles, etc. Systems which have the same configuration in one or more of these respects are said to be similar.

Similarity may be defined in two ways, by specifying the ratios either of different measurements in the same body or of corresponding measurements in different bodies. The geometrical shape of a body is determined by its intrinsic proportions, e.g., the ratio of height to breadth, and breadth to thickness, etc. In geometrically similar bodies, all such ratios (or shape factors) are constant. Alternatively, when two geometrically similar bodies are compared, there is a constant ratio between their respective heights, breadths, and other corresponding measurements which is termed the *scale ratio*. The second method has the practical advantage that a single scale ratio is substituted for a number of shape factors. For that reason, geometrical similarity is best defined in terms of correspondence and scale ratio:

but, this having been done, similarity with respect to other variables such as velocity, force, or temperature can be generally defined by a single intrinsic ratio for each system. A point-to-point geometrical correspondence between two systems ensures that if the over-all values of the intrinsic ratio are equal then corresponding point values will be equal throughout. These intrinsic ratios are the dimensionless groups which define similarity under different conditions.

Four similarity states are important in chemical engineering, namely:

Geometrical similarity
Mechanical similarity
Thermal similarity
Chemical similarity

Strictly speaking, each of the above states necessitates all the previous ones. For example, complete chemical similarity would require thermal, mechanical, and geometrical similarity. In practice, it is often necessary to accept an approximation to chemical similarity with substantial divergences from mechanical similarity.

All actual cases of similarity in fact contain an element of approximation because disturbing factors are always present which prevent ideal similarity from being attained. For example, two fluid ducts may be designed and fabricated to geometrically similar dimensions, but it is virtually impossible to make the surface roughness, for example, geometrically similar, and any differences will have some slight influence on the flow pattern. Often the effects of such departures from ideal similarity are negligible. When not negligible, they give rise to *scale effects*, and a correction of some kind has to be applied when experimental results are scaled up or down. Sometimes the requirements for similarity with respect to two important factors are totally incompatible, giving rise to the difficult case of a *mixed regime* (see Chap. 6) in which even an approximation to similarity may be impossible to achieve without drastic changes in the process.

In discussing similarity, it is necessary to refer frequently to corresponding quantities and their ratios in similar systems. Primed symbols will relate always to the large-scale systems and unprimed symbols to corresponding quantities in the model. Thus, L' would denote a given length in a large-scale or prototype apparatus, and L would be the corresponding length in a geometrically similar model. Ratios of corresponding quantities are conveniently and compactly represented by boldface type.   Thus

$$\mathbf{L} = \frac{L'}{L} = \text{linear scale ratio}$$

$$\mathbf{v} = \frac{v'}{v} = \text{ratio of corresponding velocities}$$

The numerator refers always to the large-scale system, i.e., $\mathbf{L}$ is always greater than unity irrespective of whether the model is being compared with the prototype, or vice versa.

Geometrical Similarity

Geometrical similarity, as has been said, is best defined in terms of correspondence.   Consider two solid bodies each furnished with three imaginary axes intersecting in space so that every point is uniquely described by three coordinates.   Let there be a point within the first body whose coordinates are x, y, z and a point within the second body having coordinates x', y', z', these being related to the first set of coordinates by the equation

$$\frac{x'}{x} = \frac{y'}{y} = \frac{z'}{z} = \mathbf{L}$$

where the linear scale ratio is constant.   These two points and all other pairs whose space coordinates are similarly related in terms of $\mathbf{L}$ are known as *corresponding points*.

Two bodies are geometrically similar when to every point in the one body there exists a corresponding point in the other.

The concept of geometrical similarity is illustrated in Fig. 3-1. x and x', y and y', z and z' are corresponding coordinates, P and P' are corresponding points, L and L' are corresponding lengths.

It is possible that each point in the first body may have more than one corresponding point in the second.   This occurs when the second body is composed of a multiplicity of identical elements each geometrically similar to the first body.   In this sense, a honeycomb is geometrically similar to a single dodecahedral cell.

It is not necessary that the scale ratio should be the same along each axis, and a more general definition of corresponding points is given by the equations

$$\frac{x'}{x} = \mathbf{X} \qquad \frac{y'}{y} = \mathbf{Y} \qquad \frac{z'}{z} = \mathbf{Z}$$

where **X**, **Y**, and **Z** are constant scale ratios not necessarily the same. The relation between two bodies in which scale ratios are different in different directions is one of *distorted similarity*.

The application of these geometrical concepts to process plant suggests several different kinds of small-scale apparatus that could be considered similar to a given large-scale apparatus. For convenience,

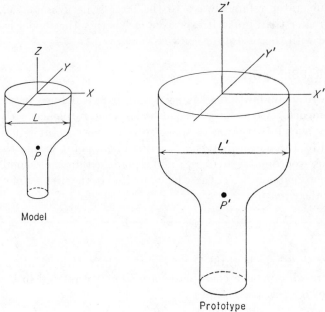

Fig. 3-1. Geometrical similarity.

the large-scale apparatus will always be referred to as the *prototype*, irrespective to whether it comes into existence first or last. A geometrically similar replica of the complete prototype on a smaller scale with equal scale ratios in all directions is termed a *model*. With different scale ratios in different directions, the small-scale apparatus becomes a *distorted model*. Where the prototype has a multiple structure composed of substantially identical elements, as, for example, a tubular heat exchanger, packed tower, filter press, electrolyzer, catalytic reactor, etc., the small-scale apparatus may be an *element*, i.e., a full-sized replica of one or more complete cells or unit components of the prototype. Or the small-scale apparatus may be a *model element*, i.e., a scale model of an element of the complete prototype. Finally, it is possible to have a *distorted model element*. These geometrical relationships are illustrated diagrammatically in Fig. 3-2.

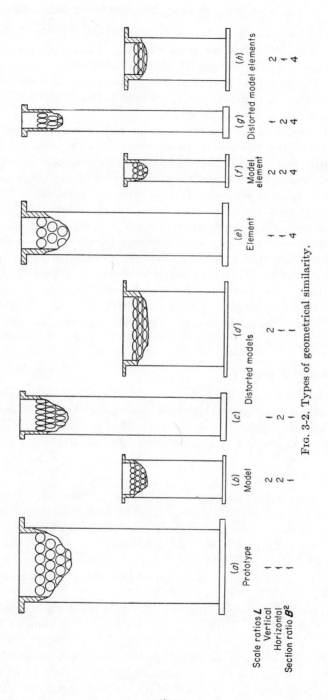

FIG. 3-2. Types of geometrical similarity.

Scale ratios L			
Vertical			
Horizontal			
Section ratio B²			

(a) Prototype
1
1
1

(b) Model
2
2
1

(c) Distorted models
1
2
1

(d) Distorted models
2
1
1

(e) Element
1
1
4

(f) Model element
2
2
4

(g) Distorted model elements
1
2
4

(h) Distorted model elements
2
1
4

17

A model is specified in terms of its *scale ratio* **L** or, in the case of a distorted model, two or more scale ratios.   An element may be specified in terms of a *section ratio* $\mathbf{B}^2$, which is the ratio of the cross-sectional area of the prototype to that of the element or of the respective numbers of unit cells or components.   A model element has both a section ratio and one or more scale ratios.

The concept of an element is useful only where any effect due to the wall of the containing vessel can be either neglected or independently controlled.   This is the case, for example, in a catalytic reactor, where the boundary surface is usually negligible compared with the interior surface.   It may even be permissible in certain circumstances to regard an empty vessel as an element of a larger one, for example, when the smaller vessel is thermally controlled by an adiabatic jacket (see Chap. 9 on Boundary Effects).   The essence of an element is that under identical conditions it shall produce the same degree of change that the prototype produces but in a smaller quantity of matter.   A packed tower is divisible vertically into elements each having the same height of packing as the prototype.   If divided horizontally or reduced in height, the parts become "differential elements," which are not directly amenable to similarity treatment.

Mechanical Similarity

Mechanical similarity comprises static, or static-force, similarity, kinematic similarity, and dynamic similarity.   Each of these can be regarded as an extension of the concept of geometrical similarity to stationary or moving systems subjected to forces.

Static Similarity

Static similarity is concerned with solid bodies or structures which are subject to constant stresses.   All solid bodies deform under stress, certain parts becoming displaced from their unstressed positions. Static similarity may be defined as follows:

Geometrically similar bodies are statically similar when under constant stress their relative deformations are such that they remain geometrically similar.

The ratio of corresponding displacements will then be equal to the linear scale ratio, and the strains at corresponding points will be the same.

In the case of elastic deformation, the condition for equality of corresponding strains is that corresponding stresses shall be in the ratio of the elastic moduli.   The ratio of the net forces acting at corresponding points in statically similar systems will be

$$\frac{F'}{F} = \mathbf{F} = \mathbf{EL}^2$$

where $\mathbf{E} = E'/E$, the ratio of the elastic moduli in prototype and model, respectively, and $\mathbf{L} =$ the linear scale ratio (prototype/model). In the case of plastic deformation, the condition for equality of corresponding strains becomes

$$\mathbf{F} = \mathbf{YL}^2$$

where $\mathbf{Y} = Y'/Y$, the ratio of the yield points of the prototype and model.

Where there is distorted geometrical similarity, the required ratios of corresponding forms for static similarity will be different in different directions. The same requirement may arise where one or both bodies are anisotropic and have different elastic moduli in different directions.

Static similarity is chiefly of interest to mechanical or structural engineers, who may employ models to predict the elastic or plastic deformation of stressed members and structures of complicated shape.

Kinematic Similarity

Kinematic similarity is concerned with solid or fluid systems in motion. Whereas geometrical similarity involved three space coordinates, kinematic similarity introduces the additional dimension of time. Times are measured from an arbitrary zero for each system, and *corresponding times* are defined as times such that $t'/t = \mathbf{t}$ is constant. $\mathbf{t}$ is the time scale ratio. Differences between pairs of corresponding times are termed corresponding intervals.

Geometrically similar particles which are centered upon corresponding points at corresponding times are termed corresponding particles.

Geometrically similar moving systems are kinematically similar when corresponding particles trace out geometrically similar paths in corresponding intervals of time.

The concept of kinematic similarity is illustrated diagrammatically in Fig. 3-3.

If the time scale ratio $\mathbf{t}$ is greater than unity, the prototype is describing corresponding movements more slowly than the model, and vice versa. The concept of a time scale ratio is less familiar than that of a linear scale ratio, and for engineering purposes it is often more convenient to calculate in terms of *corresponding velocities*, which are the velocities of corresponding particles at corresponding times. The ratio of corresponding velocities is

$$\frac{v'}{v} = \mathbf{v} = \frac{\mathbf{L}}{\mathbf{t}}$$

In the case of distorted geometrical similarity, the ratios of corresponding velocities would be different in different directions.

Kinematic similarity is a state that is of particular interest to chemical engineers because, if two geometrically similar fluid systems are kinematically similar, then the flow patterns are geometrically similar, and heat- or mass-transfer rates in the two systems will bear a simple relation to one another. Kinematic similarity in fluids entails both geometrically similar eddy systems and geometrically similar streamline boundary films. Hence, if $\mathbf{L}$ is the linear scale ratio, heat- and mass-transfer coefficients in the prototype will be $1/\mathbf{L}$ times those in the model, from which the total quantities of heat or mass transferred can easily be calculated.

In fluid systems involving, e.g., liquid jets in gases or surface waves of vortices, the flow pattern can usually be seen and measured. In closed systems with a single fluid phase, the flow pattern cannot be observed directly. The velocity at any point may, however, be measured by means of a pitot tube, and the ratio of velocities at different points is an indication of the flow pattern. For fluid flow in a full tube or cylindrical vessel, the ratio of mean to maximum velocity v/v_m is a convenient parameter. The mean velocity is obtained by dividing the cross-sectional area of the fluid path into the volumetric discharge per second: the maximum velocity can be measured by pitot tube at the axis of the tube or cylinder. For kinematically similar systems, the ratio v/v_m is constant.

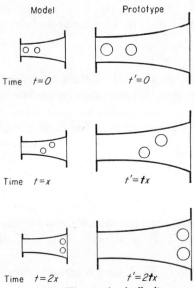

Fig. 3-3. Kinematic similarity.

Figure 3-4, which is based upon the classical researches of Stanton and Pannell, shows v/v_m for continuous flow in straight pipes, plotted against a ratio v/v_c which might be termed the *reduced velocity*. This is the mean fluid velocity divided by the lower critical velocity v_c. In the streamline region ($v/v_c < 1$) and again at high velocities, v/v_m is constant or nearly so, but immediately above the critical region it varies steeply with the velocity. Since the reduced velocity is employed instead of the actual velocity, Fig. 3-4 may be expected to apply also to coiled pipes and cylindrical vessels, although the values of v_c will be different. For channels of varying cross section or packed with rings, saddles, or granular solids, the rising curve is less steep and flattens out at a higher value of v/v_c.

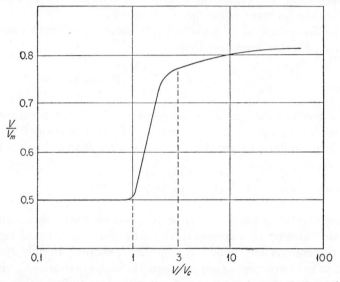

FIG. 3-4. Flow pattern in pipes. Ratio of mean to maximum velocity (v/v_m) vs. *reduced velocity* = mean velocity ÷ lower critical velocity (v/v_c). [*Based on T. E. Stanton and J. R. Pannell, Trans. Roy. Soc. (London)*, **A214**:199 (1914).]

Figure 3-4 illustrates the important conclusion that at both very low and very high velocities single-phase fluid-flow systems that are geometrically similar can be treated as kinematically similar irrespective of variations in the fluid velocity (this conclusion is derived from basic principles in Chap. 5).

Dynamic Similarity

Dynamic similarity is concerned with the forces which accelerate or retard moving masses in dynamic systems. Forces of the same kind (gravitational, centrifugal, etc.) which act upon corresponding particles at corresponding times will be referred to as *corresponding forces*.

In fluid systems or systems composed of discrete solid particles, kinematic similarity necessarily entails dynamic similarity, since the motions of the system are functions of the applied forces.   In machines or mechanical movements whose parts are constrained to follow fixed paths, it is possible to have kinematic similarity without any fixed ratio of applied forces.   In a machine, only some of the forces serve to accelerate moving masses, other forces setting up static stresses in the constraining members, overcoming frictional resistance and being dissipated as heat.

Geometrically similar moving systems are dynamically similar when the ratios of all corresponding forces are equal.

If the forces acting at a given point are of n different kinds, F_1, F_2, . . . , F_n, it is necessary that

$$\frac{F'_1}{F_1} = \frac{F'_2}{F_2} = \cdots = \frac{F'_n}{F_n} = \mathbf{F} \text{ (const)}$$

Thus the parallelograms or polygons of forces for corresponding particles will be geometrically similar.   A further consequence is that the ratios of different forces in the same system shall be constant, i.e.:

$$\frac{F'_1}{F'_2} = \frac{F_1}{F_2} \qquad \frac{F'_1}{F'_n} = \frac{F_1}{F_n} \qquad \text{etc.}$$

These are the intrinsic ratios or proportions which determine the dynamic "shape" of a system in the same manner as ratios between linear dimensions determine the geometrical shape.   In fluid systems, the principal forces acting are pressure, inertial, gravitational, viscous, and interfacial, and it is ratios between the magnitudes of these forces at corresponding points, expressed as dimensionless groups, which constitute the criteria of dynamic similarity.

For geometrically similar dynamic systems in which the physical and chemical properties of the component materials are the same (termed *homologous systems*), it is not generally possible to establish more than two ratios between three kinds of force which shall be the same in both systems.   When materials of suitably different physical properties are employed in the two systems, it becomes possible to maintain three constant ratios involving four different kinds of force.   Where the behavior of a system is significantly influenced by forces of more than four kinds, it is only in a few special cases (perhaps with the aid of geometrical distortion) that dynamic similarity can be established.

In fluid-flow systems, dynamic similarity is of direct importance where it is desired to predict pressure drops or power consumption.

In heat and mass transfer or chemical reactions, it is chiefly of indirect importance as a means of establishing kinematic similarity.

Thermal Similarity

Thermal similarity is concerned with systems in which there is a flow of heat, and it introduces the dimension of temperature as well as those of length, force (or mass), and time.

Heat can flow from one point to another by radiation, conduction, convection, and bulk movement of matter through the action of a pressure gradient. For the first three processes, a temperature difference is necessary, and, other things being equal, the rate of heat flow between any two points varies with the temperature difference between them. The fourth heat-transfer process, bulk movement of matter, depends upon the form of motion or flow pattern in the system. Hence, in moving systems thermal similarity necessitates kinematic similarity.

The temperature difference at corresponding times between a given pair of points in one system and that between the corresponding pair of points in another system will be referred to as *corresponding temperature differences.*

Geometrically similar systems are thermally similar when corresponding temperature differences bear a constant ratio to one another and when the systems, if moving, are kinematically similar.

In thermally similar systems, the patterns of temperature distribution formed by isothermal surfaces at corresponding times are geometrically similar. The ratio of corresponding temperature differences might be termed the "temperature scale ratio." Where this ratio is unity, the temperatures at corresponding points either are equal or differ from one another by a fixed number of degrees.

Thermal similarity requires that corresponding rates of heat flow shall bear a constant ratio to one another. If H_r, H_c, H_v, and H_f represent the quantities of heat transferred per second by radiation, conduction, convection, and bulk transport, respectively, across a given element of cross section, then, for thermal similarity,

$$\frac{H_r'}{H_r} = \frac{H_c'}{H_c} = \frac{H_v'}{H_v} = \frac{H_f'}{H_f} = \mathbf{H} \text{ const}$$

Alternatively, in terms of intrinsic ratios,

$$\frac{H_r'}{H_c'} = \frac{H_r}{H_c} \qquad \frac{H_r'}{H_v'} = \frac{H_r}{H_v} \qquad \frac{H_r'}{H_f'} = \frac{H_r}{H_f}$$

In general, it is not possible to maintain all three ratios simultaneously at all points, and thermal similarity can be established only when either radiation or conduction and convection are negligible.

Chemical Similarity

Chemical similarity is concerned with chemically reacting systems in which the composition varies from point to point and, in batch or cyclic processes, from instant to instant. No new fundamental dimension is introduced, but there are one or more concentration parameters, depending upon the number of independently variable chemical constituents in respect of which similarity is to be established. It is not necessary that chemical compositions in the two systems should be the same, only that there should be a fixed relation between the point concentrations of certain constituents which are being compared. Where one system contains a variable chemical constituent A and another system has a variable constituent B, and where it is desired to establish similarity with respect to A and B, these substances will be termed *corresponding constituents*.

The concentration of a given chemical constituent in an element of volume at a given time depends upon the initial concentration, the rate at which the constituent is generated or destroyed by chemical action, the rate at which it diffuses into or out of the volume element, and the rate at which it is transported by bulk movement of material. The rate of chemical action depends upon the temperature, the rate of diffusion depends upon the concentration gradient, the rate of bulk transport depends upon the flow pattern. Hence, chemical similarity necessitates both thermal and kinematic similarity, and it is dependent upon concentration differences rather than absolute concentrations. The addition of x lb/gal of constituent A which passes through the system unchanged need not alter the concentration gradients or the chemical similarity to another system containing component B.

Let the concentration difference at corresponding times between a given pair of points in one system and that between the corresponding pair of points in another system be called *corresponding concentration differences*.

Geometrically and thermally similar systems are chemically similar when corresponding concentration differences bear a constant ratio to one another and when the systems, if moving, are kinematically similar.

In chemically similar systems, the patterns formed by surfaces of constant composition at corresponding times may be thought of as being geometrically similar. The ratio of corresponding concentration differences might be called the "concentration scale ratio": where it is equal to unity, the concentrations at corresponding points either are the same or differ by a constant amount. In practice, pilot-scale chemical reactors are nearly always operated under the same temperature and concentration conditions as the prototype, and the similarity relation aimed at is that of equal temperatures and product concentrations at corresponding points and times, which is a special case under the general definition given above.

The intrinsic ratios or criteria which define chemical similarity, in addition to those required for kinematic and thermal similarity, are

$$\frac{\text{Rate of chemical formation}}{\text{Rate of bulk flow}}$$

and

$$\frac{\text{Rate of chemical formation}}{\text{Rate of molecular diffusion}}$$

The second ratio can often be neglected by comparison with the first.

In theory, the rate of a chemical reaction may be independently varied by changing the temperature. In practice, both the chemical equilibrium and the relative rates of unwanted side reactions vary with temperature, and there is usually a rather narrow temperature range within which the reaction must proceed on both the small and the large scale in order to ensure maximum yield. In both model and prototype, the reaction time will be of the same order, and this requirement fixes the relative velocities in continuous-flow systems. These velocities are incompatible with the velocities necessary for kinematic similarity except at very low or very high velocities, as was illustrated by Fig. 3-4. Hence, in scaling up a continuous chemical reaction, and especially where there is an optimum reaction time beyond which the yield or quality is reduced, it is advantageous to operate both model and prototype either in the streamline region or with a high degree of turbulence. If neither condition is feasible, there will be an unpredictable scale effect and it would be prudent either to scale up in several stages or to allow ample factors of safety in design.

Similarity Criteria

It has been mentioned that mechanical, thermal, or chemical similarity between geometrically similar systems can be specified in terms of criteria which are intrinsic ratios of measurements, forces, or rates within each system. Since these criteria are ratios of like quantities, they are dimensionless and there are two general methods of arriving at them.   Where the differential equations that govern the behavior of the system are unknown, but provided we do know all the variables which would enter into the differential equations, it is possible to derive the similarity criteria by means of dimensional analysis. Where the differential equations of the system are known but cannot be integrated, the similarity criteria can be derived from the differential form.   Where the differential equations are both known and capable of being integrated and solved, there is in general no need for either similarity criteria or model experiments, since the behavior of a large-scale system should be directly calculable.

SYMBOLS IN CHAPTER 3

B^2 = section ratio
E = elastic modulus
F = force
H = heat flux
L = linear dimension
$\mathbf{L}$ = scale ratio
t = time
v = mean velocity
v_c = lower critical velocity
v_m = maximum velocity
x, y, z = linear coordinates at right angles
Y = yield point

CHAPTER 4

DIMENSIONAL ANALYSIS

Dimensional analysis is a technique for expressing the behavior of a physical system in terms of the minimum number of independent variables and in a form that is unaffected by changes in the magnitude of the units of measurement. The physical quantities are arranged in dimensionless groups consisting of ratios of like quantities—lengths, velocities, forces, etc.—which characterize the system. These groups constitute the variables in the dimensionless equation of state (or motion) of the system.

Dimensional analysis can give misleading results unless every variable which significantly influences the system is taken into account. A good deal must therefore be known about the mechanism of a process before the dimensional method can be applied with confidence. The crux of the matter lies in the initial choice of variables. If the list is too long and includes variables whose effect is negligible, the superfluous factors are sometimes shed in the course of the analysis, but in other cases the number of derived similarity criteria becomes unnecessarily large, and the problem of establishing similarity is made to appear more difficult than it really is. If, on the other hand, any single relevant variable is omitted, dimensional analysis will lead to a false conclusion. As a further complication there are the so-called *dimensional constants*, the need for which may not be immediately obvious, but which must be taken into account if the analysis is to lead to a correct result. For these reasons, dimensional analysis alone has seldom led to completely new knowledge. Most of the application quoted by Rayleigh, Bridgman, and others amount to the confirmation of relations which were already known or suspected, so that the main difficulty of the initial selection of variables did not arise.

Units and Dimensions

Physical quantities are measured in units of different kinds, depending upon the nature of the physical quantity and the operations performed in measuring it. The *units* of measurement in any system are

27

by definition interrelated.   For example, a unit of velocity is related to units of length and time, a unit of viscosity is related to units of force, velocity, and length, and so on.   All systems of measurement thus have the property that after a small number of units have been given arbitrary values the absolute magnitudes of the other units are determined.   The number of units which must be fixed in order to determine the rest depends upon the kind of physical system that is to be measured.   In purely dynamic systems, it is sufficient to fix the size of units for any three independent variables in order to determine all the other units that are needed in dynamical measurements.   The independent variables selected are usually either mass, length, and time (m, L, t) or force, length, and time (F, L, t), and these are termed *fundamental*, or *primary*, quantities.   The units in which primary quantities are measured will be referred to as *primary units*.   It is theoretically possible to choose any other three independent variables as primary quantities, for example, stress, area, and velocity.   The unit of time would then be a secondary unit, its dimension in terms of primary units being $\sqrt{\text{area}}/\text{velocity}$.

The measurement of systems in which there are temperature changes requires a fourth independent unit to be defined before the values of all the thermal units are fixed.   The fourth primary quantity is usually taken as temperature (T).   Electromagnetic systems also require an extra primary quantity, which may be either electric charge or magnetic induction.   It will be shown that in dimensional analysis it is sometimes advantageous to increase the number of primary quantities beyond the minimum number required to fix the other units.   Thus, in analyzing thermal systems, heat (Q) is often taken as a primary quantity in addition to temperature.

There is one important condition to which all units of measurement employed in dimensional analysis are required to conform.   The ratios of the numbers denoting two different values of a given variable must remain the same when the absolute magnitude of the unit is changed.   For example, the ratio of two temperatures on the absolute scale is the same whether they are measured in degrees Kelvin or degrees Rankine, but this does not hold for degrees Centigrade and degrees Fahrenheit; hence, the latter units are inadmissible in dimensional analysis.   Similarly, density may be expressed in pounds per cubic foot, pounds per gallon, or grams per cubic centimeter, but not in degrees Twaddell or API.   This is termed *the condition of the absolute significance of relative magnitude*.

Bridgman proved that, when a set of primary units fulfills the condition of the absolute significance of relative magnitude, and only

then, all secondary units of measurement can be expressed as a product
of powers of the primary units multiplied by a constant (246). For
example, if the absolute magnitudes of the primary units of mass,
length, and time are represented by the symbols m, L, t, the magnitude
of the corresponding unit of viscosity is given by the expression

$$C \left[\frac{m}{Lt} \right]$$

where C is a constant. This expression shows at once that, if the absolute magnitude of the primary units were to be increased to $2m$, $3L$,
and $4t$, respectively, the unit of viscosity would be

$$\left[\frac{2}{3 \times 4} \right]$$

or $\frac{1}{6}$ times the size of the former unit. Hence, the number denoting
a given viscosity in the new system would be 6 times that in the old
system. The particular product of powers of the primary units
required to express the unit of measurement of a secondary quantity is
termed the *dimensional formula*, or simply the *dimensions*, of the
secondary quantity. Dimensional formulas are customarily enclosed
in brackets, and the constant C is omitted. Thus, the dimensions of
viscosity in the mLt system are $[m/Lt]$.

The British system of consistent units as used in static and dynamic
calculations belongs to the mLt class and is founded on the pound mass,
foot, and second (or hour). The unit of force is the poundal, defined
in accordance with Newton's law of inertia as the force required to
impart to a mass of one pound an acceleration of one foot per second.
A force of 1 lb weight is equal to about 32.2 poundals, depending on the
local value of the acceleration of gravity. The basic American system
of consistent units, on the other hand, belongs to the FLt class and is
based upon the pound force, foot, and second (or hour). The pound
force is defined as equal to a pound weight under standard conditions,
i.e., when the acceleration of gravity is 32.1740 ft/sec^2 (the "normal"
value adopted by the International Bureau of Weights and Measures
in 1892). To preserve consistency with Newton's law, a new unit of
mass is chosen called the *slug*, defined as the mass to which a force of
one pound would impart an acceleration of one foot per second. A
slug is equal to exactly 32.1740 lb mass.

In both of the above systems, the units are defined so that force is
numerically equal to mass times acceleration. Engineers generally
use a more complex system of units in which the dimensions of both

mass and force are accepted as basic and are measured, for example, in mass pounds and force pounds, respectively. This usage required a constant to be introduced into Newton's law of inertia, which becomes

$$Fg_c = ma$$

where g_c is a constant having the dimensions $\dfrac{m}{F}\dfrac{L}{t^2}$ and numerically

TABLE 4-1. DIMENSIONS OF MECHANICAL QUANTITIES

Quantity	Symbol	Dimensions		
		mLt	FLt	$FmLt$
Velocity..................	v	$\dfrac{L}{t}$	$\dfrac{L}{t}$	$\dfrac{L}{t}$
Acceleration...............	a	$\dfrac{L}{t^2}$	$\dfrac{L}{t^2}$	$\dfrac{L}{t^2}$
Mass......................	m	m	$\dfrac{Ft^2}{L}$	m
Force.....................	F	$\dfrac{mL}{t^2}$	F	F
Conversion factor............	g_c	$\cdots$	$\cdots$	$\dfrac{m}{F}\dfrac{L}{t^2}$
Weight....................	W	$\dfrac{mL}{t^2}$	F	F
Momentum.................	mv	$\dfrac{mL}{t}$	Ft	Ft
Energy....................	E	$\dfrac{mL^2}{t^2}$	FL	FL
Power.....................	P	$\dfrac{mL^2}{t^3}$	$\dfrac{FL}{t}$	$\dfrac{FL}{t}$
Density...................	ρ	$\dfrac{m}{L^3}$	$\dfrac{Ft^2}{L^4}$	$\dfrac{m}{L^3}$
Stress..................... Pressure...................	$\left.\begin{matrix}S\\p\end{matrix}\right\}$	$\dfrac{m}{Lt^2}$	$\dfrac{F}{L^2}$	$\dfrac{F}{L^2}$
Viscosity..................	μ	$\dfrac{m}{Lt}$	$\dfrac{Ft}{L^2}$	$\dfrac{Ft}{L^2}$

equal to the acceleration of gravity under standard conditions, that is, 32.1740 if m and F are both measured in pounds. A similar practice prevails in metric countries. In the cgs system, the gram and dyne are defined consistently with Newton's law, $F = ma$. Continental engineers nevertheless employ the kilogram as both a unit of mass and a unit of force so that a constant g_c numerically equal to 981 $\left(\dfrac{\text{kg mass} \times \text{cm}}{\text{kg force} \times \text{sec}^2}\right)$ must be introduced. The Newton's-law constant

g_c is one of the so-called *dimensional constants*, of which more will be said later.

Table 4-1 gives the dimensional formulas of certain mechanical quantities in the mLt, FLt, and $FmLt$ systems of dimensions. The formulas are shown in fractional form for the sake of clearness, but in order to save space they are usually written on one line, using negative exponents where necessary, thus:

$$[mL^{-1}t^{-1}]$$

Dimensional formulas are useful for converting a measurement from one system of units to another. Suppose it is required to convert a viscosity from cgs units (poises) to foot-pound-hour units. The foot-pound-hour unit will be

$$\frac{454}{30.48 \times 3,600}$$

or 0.00416 times the size of the cgs unit; hence, the cgs measurement must be multiplied by $1/0.00416 = 242$ to give the foot-pound-hour figure.

Dimensional Homogeneity

In representing a physical process by means of a mathematical equation, we tacitly assume that the separate terms of the equation stand for physical quantities that can significantly be added together. For example, it is arithmetically possible to add together pounds avoirdupois and pounds sterling, but the figure so obtained would have no practical value or meaning. For the mathematical operations of addition or subtraction to have any physical significance, it is necessary that the terms added or subtracted should denote physical quantities of the same kind; in other words, every term in a physical equation should have the same dimensional formula, or otherwise the equation is meaningless. This is called the *principle of dimensional homogeneity*.

The principle of dimensional homogeneity is sometimes loosely stated in some such words as that all complete physical equations are dimensionally homogeneous. Bridgman pointed out that this is not necessarily true and gave the following example (246): Consider the standard dynamic equations for the velocity v and distance s traversed by a falling body:

$$v = gt$$
$$s = \tfrac{1}{2}gt^2$$

Simple addition of the two equations gives

$$v + s = gt + \tfrac{1}{2}gt^2$$

This is formally a complete and valid equation; yet it is not dimensionally homogeneous. Two of the terms have the dimensions $[Lt^{-1}]$ and two the dimensions $[L]$. Such an equation is, of course, physically meaningless. No practical significance can be attached to the sum of a velocity and a distance.

The principle of dimensional homogeneity is correctly stated as follows:

Every complete physical equation is either dimensionally homogeneous or capable of being resolved into two or more separate equations that are dimensionally homogeneous.

A *complete* physical equation is one which remains valid when the units of measurement are changed. In many empirical equations, the values of the numerical constants change when the units are changed. This shows that all the relevant variables do not appear explicitly in the equations, some of them being concealed in the numerical "constants." Such equations are not complete.

The principle of dimensional homogeneity is useful both in checking theoretical derivations and in guiding empirical correlations. If a derived or assumed equation is not dimensionally homogeneous, then an error has been made, some relevant factor has been neglected, or the equation has not been reduced to its simplest terms. An equation is checked for dimensional homogeneity by substituting for each physical quantity its dimensional formula (numerical constants being omitted). Thus the dynamical equation

$$s = vt + \tfrac{1}{2}at^2$$

(where a = acceleration, other symbols as above) becomes dimensionally

$$[L] = [Lt^{-1}][t] + [Lt^{-2}][t^2]$$

It is seen at once that every term has the dimensions $[L]$, and hence the equation is dimensionally homogeneous.

The quantities which enter into physical equations are of three kinds: physical variables, numerical constants, and dimensional constants. The dimensions and dimensional formulas of physical variables were discussed in the previous section. Numerical constants have "zero" dimensions, i.e., the exponent of each of the primary quantities in the dimensional formulas is zero, so that the dimensional formula reduces to [1]. Hence, the value of a numerical constant does not change when the units of measurement are changed. An example of a numerical constant is the quantity $\pi = 3.14159 \cdots$.

A dimensional constant is any constant appearing in a physical equation which changes in value when the units of measurement of the primary quantities are changed.

Dimensional constants have dimensional formulas like those of physical variables, by means of which they can be transformed from one system of units into another. Mention has already been made of the Newton's-law constant $g_c = 32.1740$, which has the dimensions $[mF^{-1}Lt^{-2}]$. A dimensional constant may be regarded as a kind of conversion factor introduced into an equation in which inconsistent units are employed, e.g., the engineering units of force and mass. A dimensional constant can always be eliminated by redefining the units of measurement so as to make the constant equal to unity, but it is often more convenient to employ the customary units and introduce the appropriate dimensional constant. Other dimensional constants which are met with in chemical engineering include the gas constant R, the mechanical equivalent of heat J, and the various chemical-reaction velocity constants.

The dimensional constants which must be introduced into the dimensional analysis of a given system or process are those which would appear in the equations of motion of the system, or the rate equations for the process, if these were written down in full. The principle of dimensional homogeneity presupposes that the dimensions of dimensional constants have been taken into consideration as well as those of physical variables; hence, if a theoretically derived equation proves to be nonhomogeneous, one of the first steps is to check whether some necessary dimensional constant has been omitted.

Any dimensionally homogeneous physical equation can be expressed as a zero function of dimensionless groups of variables by dividing throughout by any one term and rearranging. For example, the dynamical equation

$$s = vt + \tfrac{1}{2}at^2$$

can be written as either

$$\frac{vt}{s} + \frac{at^2}{2s} - 1 = 0$$

or

$$\frac{s}{vt} - \frac{at}{2v} - 1 = 0$$

or

$$\frac{2s}{at^2} - \frac{2v}{at} - 1 = 0$$

Buckingham's Theorem

At the beginning of the chapter, it was said that dimensional analysis is a technique for arranging a given set of physical variables in the maximum number of independent dimensionless groups. This statement can now be amplified. *A complete set of dimensionless groups* is defined as any set derived from a given collection of variables and dimensional constants and such that, first, every group in the set is independent of the others and, second, any other possible dimensionless group from the same variables and constants can be obtained as a product of powers of groups in the complete set. A complete set thus contains the maximum number of independent dimensional groups that can be derived from a given collection of variables and dimensional constants.

It has been shown that every dimensionally homogeneous equation can be expressed as a zero function of a set of dimensionless groups. Buckingham (247, 248) stated further that every such equation can be expressed as a zero function of a *complete* set of dimensionless groups. This important generalization is known as *Buckingham's theorem,* sometimes called the II theorem.

It falls into two parts which may be stated as follows:

1. *The solution to every dimensionally homogeneous physical equation has the form*

$$\phi(\Pi_1, \Pi_2, \ldots) = 0$$

where Π_1, Π_2, . . . *represent a complete set of dimensionless groups of the variables and dimensional constants in the equation.*

2. *If an equation contains n separate variables and dimensional constants and these are given dimensional formulas in terms of m primary quantities, then the number of dimensionless groups in a complete set is* $n - m$.

From part 1 of Buckingham's theorem, it follows that, where the physical equation relating a given collection of variables is unknown, the arguments Π, Π_2, . . . of the equation in its dimensionless form can be found by constructing a complete set of dimensionless groups from the variables and any dimensional constants that may be involved. The form of the function ϕ and the values of any numerical constants remain undetermined, but it is often unnecessary to know these in order to establish conditions of similarity.

Part 2 of Buckingham's theorem shows that, the larger the number of primary quantities m can be made without causing an increase in n, the smaller the number of dimensionless groups in the complete set, and the more specific the information which dimensional analysis can give. This is the reason why heat is often taken as a primary quantity in addition to mass (or force), length, and time, in systems where there is no appreciable conversion of heat into mechanical energy, or vice versa. If there were conversion of heat into mechanical energy, it would be necessary to introduce a dimensional constant, the mechanical equivalent of heat J. Both n and m would then be increased by 1, and no advantage would be gained from the additional primary quantity. Similarly, it was shown in the previous section that when both mass and force are taken as primary quantities (as in the ordinary engineering system or units) the Newton's-law constant g_c appears in all equations embodying both mass and force, so that again the additional primary quantity leads to no reduction in the total number of dimensionless groups.

Part 1 of Buckingham's theorem is always true, but there are occasional exceptions to part 2. If it should happen that when the dimensional formulas of all the variables are written down two of the primary quantities appear constantly in the same relation, so that they could in every case be replaced by a single symbol, then the number of dimensionless groups in a complete set will be one more than predicted by the difference $n - m$. In effect, the number of *independent* primary quantities is reduced to $m - 1$, since two of them are related, and hence the number of dimensionless groups becomes $n - (m - 1)$. A strict algebraic rule for the number of independent dimensionless groups in a complete set, based upon the theory of determinants, will be mentioned later. In practice, the exceptions to part 2 of Buckingham's theorem are found without difficulty when the method of indices is applied as described in the next section.

Rayleigh's Method of Indices

The usual method of deriving a complete set of dimensionless groups from a given collection of physical variables and dimensional constants is the *method of indices*, an algebraic procedure which was first employed by Rayleigh (254). The method is illustrated by the following example: Suppose it is desired to obtain a complete set of dimensionless groups from the physical variables which influence the isothermal flow of a fluid and that the mLt system of primary quantities is to be employed. The variables with their symbols and dimensional formulas are as follows:

Fluid velocity	v	$[Lt^{-1}]$
Linear dimension	L	$[L]$
Force	F	$[mLt^{-2}]$
Density	ρ	$[mL^{-3}]$
Viscosity	μ	$[mL^{-1}t^{-1}]$
Surface tension	σ	$[mt^{-2}]$
Acceleration of gravity	g	$[Lt^{-2}]$

Let it be stipulated that a consistent unit of force is to be employed (i.e., consistent with the unit of mass and Newton's law) so that no dimensional constant is involved. There are seven variables and three primary quantities; hence, Buckingham's theorem (part 2) would lead one to expect a set of four independent dimensionless groups.

If all of the above variables influence the isothermal flow of a fluid, then the equation of motion of such a system must be capable of being expressed in the form

$$\phi(V,L,F,\rho,\mu,\sigma,g) = 0$$

Whatever the unknown function ϕ, the equation can be rearranged so that any one variable enters directly. It is convenient to make this the dependent variable F and rewrite the equation:

$$\phi(V,L,\rho,\mu,\sigma,g)F = 1$$

Provided this equation is dimensionally homogeneous, and provided the units of measurement fulfill the condition of the absolute significance of relative magnitude, the dimensional formulas of the variables must conform to a power relation

$$[v]^{\alpha}[L]^{\beta}[\rho]^{\gamma}[\mu]^{\delta}[\sigma]^{\epsilon}[g]^{\theta}[F] = 0$$

where $\alpha, \beta, \gamma, \ldots$ represent unknown indices and the square brackets round the symbols indicate that the dimensional formulas of the variables are referred to and not the variables themselves. Substituting the dimensional formulas,

$$[Lt^{-1}]^{\alpha}[L]^{\beta}[mL^{-3}]^{\gamma}[mL^{-1}t^{-1}]^{\delta}[mt^{-2}]^{\epsilon}[Lt^{-2}]^{\theta}$$
$$[mLt^{-2}] = 0$$

Equating indices of the primary quantities m, L, and t results in three simultaneous equations, known as equations of condition,

Condition on m: $\gamma + \delta + \epsilon + 1 = 0$
Condition on L: $\alpha + \beta - 3\gamma - \delta + \theta + 1 = 0$
Condition on t: $-\alpha - \delta - 2\epsilon - 2\theta - 2 = 0$

These equations can be solved for any three of the unknown indices giving a set of solutions in terms of the other three. Depending upon which indices are eliminated, different sets of dimensionless groups will be obtained. Solving in terms of δ, ϵ, and θ,

$$\alpha = -\delta - 2\epsilon - 2\theta - 2$$
$$\beta = -\delta - \epsilon + \theta - 2$$
$$\gamma = -\delta - \epsilon - 1$$

whence

$$[v]^{-\delta-2\epsilon-2\theta-2}[L]^{-\delta-\epsilon+\theta-2}[\rho]^{-\delta-\epsilon-1}[\mu]^{\delta}[\sigma]^{\epsilon}[g]^{\theta}[F] = 0$$

or

$$\left[\frac{vL\rho}{\mu}\right]^{-\delta} \left[\frac{\rho v^2 L}{\sigma}\right]^{-\epsilon} \left[\frac{v^2}{Lg}\right]^{-\theta} \left[\frac{F}{\rho v^2 L^2}\right] = 0$$

Therefore, by Buckingham's theory, part 1, the equation of motion of an isothermal fluid flow must have the form

$$\phi\left(\frac{vL\rho}{\mu}, \frac{\rho v^2 L}{\sigma}, \frac{v^2}{Lg}\right) \frac{F}{\rho v^2 L^2} = 1$$

or

$$\frac{F}{\rho v^2 L^2} = \phi\left(\frac{vL\rho}{\mu}, \frac{\rho v^2 L}{\sigma}, \frac{v^2}{Lg}\right)$$

where the function ϕ is undetermined.

The above four groups constitute a complete set of dimensionless groups from the variables listed at the beginning of the example. These groups are all well known in connection with fluid dynamics and are given the following names:

$$\frac{F}{\rho v^2 L^2} \ldots \ldots \text{Pressure coefficient}$$

$$\frac{vL\rho}{\mu} \ldots \ldots \text{Reynolds number}$$

$$\frac{\rho v^2 L}{\sigma} \ldots \ldots \text{Weber group}$$

$$\frac{v^2}{Lg} \ldots \ldots \text{Froude group}$$

By solving the condition equations in terms of other indices, say, α, β, and γ, a different set of dimensionless groups would have been obtained. Any complete set of dimensionless groups from given variables can, however, be transformed into any other complete set by combination and rearrangement.

In the above solutions, the variables corresponding to the unknown indices δ, ϵ, and θ—that is, μ, σ, and g—as well as the variable F, whose index was unity, occur each in only one dimensionless group. In deriving dimensionless equations that are to be verified experimentally, it is desirable that those variables which are most easily controlled, as well as the dependent variable itself (in this case, F), should each appear in only one dimensionless group. This will be achieved if the index of the dependent variable is made equal to unity and the condition equations are solved in terms of indices corresponding to easily controllable independent variables.

A convenient method of carrying out this and other types of dimensional calculation is to set out the dimensional formulas of the variables in tabular form, thus:

Variable:	v	L	ρ	μ	σ	g	F
Index:	α	β	γ	δ	ϵ	θ	1
Exponent of m.......	...	...	1	1	1	...	1
Exponent of L.......	1	1	-3	-1	...	1	1
Exponent of t........	-1	...	...	-1	-2	-2	-2

Such a table is called a *dimensional matrix*, and it shows directly the numerical coefficients of the indices α, β, γ, . . . in the equations of condition.

The dimensional matrix forms the basis of a rigorous algebraic treatment of dimensional equations which was first employed by Bridgman (246) and which has been developed and extended by Van Driest (260) and Langhaar (251). Bridgman pointed out that if a dimensional matrix is written in terms of m primary quantities there will be more independent dimensionless groups than are predicted by Buckingham's theorem in the special case where each of the m-rowed determinants formed out of the matrix is equal to zero. This is a formal way of stating the condition referred to in the previous section, in which one of the primary quantities either does not enter any of the dimensional formulas or bears a fixed relation to another primary quantity. In the first case, there would be only two condition equations instead of three; in the second case, two of the condition equations would be found to give the same information. In either case, the result is one more unknown index and an extra dimensionless group.

The rigorous rule corresponding to part 2 of Buckingham's theorem may be enunciated as follows:

The number of dimensionless groups in a complete set is equal to the total number of variables plus dimensional constants less the numbers of rows in the highest nonzero determinant that can be formed out of the dimensional matrix.

Langhaar (251) has described a systematic procedure for deriving a complete set of dimensionless groups which is based on the use of determinants and which provides for the occasional exception to part 2 of Buckingham's theorem. In practice, however, these exceptions cause no trouble when the simpler method of indices is used. From the equations of condition, it is easily seen whether there will be more than the normal number of unknowns.

Primary Quantities

As has already been mentioned, the choice of primary quantities in dimensional analysis is largely a matter of convenience. For the problems of statics and stress analysis, two primary quantities are sufficient, force and length. There is no motion, and hence time does not enter. The dynamics of solid particles and isothermal flow of fluids requires three primary quantities, and either force, length, and time or mass, length, and time are commonly chosen. Temperature on the absolute thermodynamic scale is defined as proportional to the kinetic energy of the atom, i.e., to energy content per mole divided by Avogadro's number. Hence, in either of the above three-dimensional systems, temperature could be given the dimensions of energy. However, it is usually convenient to take temperature as a fourth primary quantity and provide for the relation with kinetic energy, when it enters, by means of a dimensional constant, the *gas constant R*. This leads to a four-dimensional system whose primary quantities are either force, length, time, and temperature (F, L, t, T) or mass, length, time, and temperature (m, L, t, T).

Ordinary engineering practice employs independent units of force and mass, namely, the pound (force) and the pound (mass), and also independent units of heat and energy, the Btu and foot-pound. There is no objection to employing these units in dimensional analysis provided that the appropriate dimensional constants are introduced when interaction occurs, namely, the Newton's-law constant g_c in respect of the units of force and mass, and the mechanical equivalent of heat J in respect to the units of heat and energy. The use of engineering units is equivalent to employing a six-dimensional system whose primary quantities are force, mass, length, time, temperature, and quantity of heat (F, m, L, t, T, Q). In cases where the dimensional constants do not enter into the equations of motion, the six-dimensional system has an advantage over the four-dimensional system in that it leads to a smaller number of independent dimensionless groups and therefore gives more specific information about the process under analysis. McAdams (161) has recommended the six-dimensional system for the dimensional analysis of heat-transfer processes. In a recent work (250), Huntley proposes to increase the number of primary quantities even further by treating lengths as vectors and introducing three linear dimensions instead of one. This was originally suggested by Williams (261) and has certain theoretical advantages, but it has not been widely adopted.

Appendix 2 gives the dimensions of the principal physical and chemical quantities and dimensional constants in both the $FmLtTQ$ (six-dimensional) and the $mLtT$ (four-dimensional) systems.

As a final example, the method of indices will be applied to the transfer of heat by natural convection, a process that it is difficult to analyze mathematically. The variables are:

Heat transferred in unit time...........	H	$[Qt^{-1}]$
Temperature difference.................	ΔT	$[T]$
Linear dimension.....................	L	$[L]$
Fluid density.........................	ρ	$[mL^{-3}]$
Viscosity (mass basis).................	μ	$[mL^{-1}t^{-1}]$
Specific heat.........................	c	$[Qm^{-1}T^{-1}]$
Thermal conductivity.................	k	$[QL^{-1}t^{-1}T^{-1}]$
Coefficient of cubical expansion.........	β	$[T^{-1}]$
Acceleration of gravity................	g	$[Lt^{-2}]$

There are nine variables and only five primary quantities, since force does not appear in any of the dimensional formulas. From Buckingham's theorem, one would expect four dimensionless groups. This, however, is a case where some knowledge of the mechanism of the process enables the number of independent variables to be reduced. Natural convection is due to buoyancy forces which are proportional to the product βg, and it is only in relation to buoyancy that gravity significantly affects the process. Hence, instead of taking β and g separately, we may substitute the product βg, whose dimensions are $[Lt^{-2}T^{-1}]$. This reduces the number of variables to eight.

Writing down the dimensional power equation,

$$[H]^\alpha[\Delta T]^\beta[L]^\gamma[\rho]^\delta[\mu]^\epsilon[c]^\theta[k]^\iota[\beta g]^\kappa = 0$$

Expressed as a dimensional matrix,

Variable: Exponent:	H α	ΔT β	L γ	ρ δ	μ ϵ	c θ	k ι	βg κ
Power of m..	...	...	...	1	1	−1		
Power of L..	...	...	1	−3	−1	...	−1	1
Power of t...	−1	...	...	...	−1	...	−1	−2
Power of T..	...	1	...	...	...	−1	−1	−1
Power of Q..	1	...	...	...	...	1	1	

The equations of condition are therefore:

In m: $$\delta + \epsilon - \theta = 0$$
In L: $$\gamma - 3\delta - \epsilon - \iota + \kappa = 0$$
In t: $$-\alpha - \epsilon - \iota - 2\kappa = 0$$
In T: $$\beta - \theta - \iota - \kappa = 0$$
In Q: $$\alpha + \theta + \iota = 0$$

Solving in terms of α, θ, and κ,

$$\beta = \kappa - \alpha$$
$$\gamma = 3\kappa - \alpha$$
$$\delta = 2\kappa$$
$$\epsilon = \theta - 2\kappa$$
$$\iota = -\alpha - \theta$$

whence

$$\left[\frac{H}{\kappa L\,\Delta T}\right]^{\alpha}\left[\frac{\beta g\,\Delta T\,L^3\rho^2}{\mu^2}\right]^{\kappa}\left[\frac{c\mu}{k}\right]^{\theta} = 1$$

or

$$\frac{H}{kL\,\Delta T} = \phi\left(\frac{\beta g\,\Delta T\,L^3\rho^2}{\mu^2},\frac{c\mu}{k}\right)$$

In terms of the heat-transfer coefficient $h = H/L^2\,\Delta T$,

$$\frac{hL}{k} = \phi\left(\frac{\beta g\,\Delta T\,L^3\rho^2}{\mu^2},\frac{c\mu}{k}\right)$$

The first group on the right is known as the *Grashof number:* it represents the ratio of buoyancy to viscous forces in the system multiplied by the ratio of inertial to viscous forces (hence the entry of the viscosity as the square). The second group on the right is called the *Prandtl group* and represents the ratio of kinematic viscosity (or momentum diffusivity) to thermal diffusivity.

If the viscosity had been taken on a force basis instead of a mass basis, it would have been necessary to introduce a dimensional constant, the Newton's-law constant g_c. This would have increased the number of variables to 10, but the number of primary quantities would simultaneously have been increased to 6 by the appearance of force. Again, had we not used the trick of combining β and g in a single variable, the result of the dimensional analysis would have been different and much less useful (the student might work this out). These points illustrate some of the pitfalls associated with dimensional analysis which limit the practical utility of the method in the absence of analytical knowledge.

SYMBOLS IN CHAPTER 4

a = acceleration
c = specific heat
E = energy
F = force
g = acceleration of gravity
g_c = Newton's-law constant
H = heat flux
J = mechanical equivalent of heat
k = thermal conductivity

L = length
m = mass
m = number of primary quantities
n = number of variables
P = power
p = pressure
Q = heat
R = gas constant
S = stress
s = distance traveled
T = temperature
t = time
v = velocity
W = weight
β = coefficient of cubical expansion of liquid
Δ = difference
μ = viscosity
ρ = density
σ = surface tension
ϕ = function
Π = dimensionless group

CHAPTER 5

DIFFERENTIAL EQUATIONS

For most of the physical and chemical processes that are utilized in chemical engineering, the fundamental differential equations are known: the difficulty is to integrate them. If this could be done mathematically for systems of every shape and degree of complexity, then pilot plants and model experiments would become unnecessary, since the behavior of any large-scale system could be predicted on the basis of laboratory data alone. For single processes occurring in systems of simple geometrical shape, it sometimes is possible to integrate mathematically, as for the streamline flow of fluids in straight pipes (leading to Poiseuille's equation). Very often, however, the system or the process is so complex that a mathematical integration is impossible, and it is necessary to integrate empirically. Most of the standard rate equations of chemical engineering are in effect empirical integrations of the fundamental differential equations; as such, they are strictly valid only where the system to which they are applied bears some geometrical resemblance to the systems from which they were experimentally derived.

Where the differential equation governing a particular process is known, and provided that it is complete and dimensionally homogeneous (see Chap. 4), it is easy to put the equation into dimensionless form and thus derive the similarity criteria without the aid of dimensional analysis. The procedure is first to reduce the differential equation to a generalized dimensional form, omitting differential signs and numerical constants, then divide through by any one term to render all the terms dimensionless (care must be taken *not* to omit dimensional constants, or the generalized form may not be dimensionally homogeneous). Examples appear later in this chapter, e.g., Eqs. (5-10) to (5-13), (5-15) to (5-17), etc.

There are advantages in deriving the similarity criteria from differential equations rather than by a dimensional analysis of the original variables. In the first place, the former method overcomes the major

43

difficulty of dimensional analysis, that of ensuring that no significant variables have been neglected. Second it shows up the physical significance of dimensionless criteria as ratios of flows, forces, or analogous quantities. The information obtained by this method is of course subject to any errors that may exist in the differential equations, e.g., through neglecting minor variables: but it avoids the additional errors due to incorrect integration or inaccurate constants. A model experiment based upon similarity criteria which have been derived from the fundamental differential equations effects an empirical integration of the equations for the particular system and geometry that are being studied. If there should be any variables neglected in the differential equation which are found to influence the process significantly, dimensional analysis can be applied directly to such variables in order to supplement the information obtained from the differential equations. For example, the Navier-Stokes equations for fluid flow, when treated as described above, yield a dimensionless equation which represents the pressure coefficient as a function of the Reynolds number and Froude group. A dimensional analysis of surface-tension effects enables the Weber group to be added [see Eqs. (5–25)ff.].

Before the derivation of similarity criteria is discussed in detail, it will be useful to consider some general aspects of physical and chemical rate equations, an appreciation of which will help in the interpretation and correct use of the criteria.

Every physical equation, whether differential or integral, can be resolved into three factors, which constitute, respectively:

1. A driving force or difference in potential
2. A resistance factor or its reciprocal, a conductance
3. A resultant

Items 1 and 2 are chosen so that the resultant is directly proportional to the driving force and inversely proportional to the resistance. The driving force of a process or reaction does not necessarily have the dimensions of a force in the mechanical sense; it may, for example, be a difference in temperature or concentration tending to produce a flow of heat or matter. This way of looking at physical processes is sometimes referred to as the "potential concept." The classical example is Ohm's law in electricity,

$$I = \frac{E}{R_E} \tag{5-1}$$

where I = current
  E = applied voltage
  R_E = electrical resistance

Similarly, one can write for heat transfer by conduction

$$H = \frac{\theta}{R} \tag{5-2}$$

where H = rate of flow of heat (resultant)

θ = temperature difference (driving force)

R = thermal resistance

The electrical analogy extends to resistances in series and parallel. In some physical processes, the resultant depends upon the magnitudes of several reaction resistances r_1, r_2, r_3, . . . in series. Alternatively, the total driving force may be divided into separate parts θ_1, θ_2, θ_3 representing the potential difference (e.g., temperature drop) across each series resistance. The potential equation is then written in the *series form*, either

$$H = \frac{\theta}{r_1 + r_2 + r_3 + \cdots} \tag{5-3}$$

or

$$H = \frac{\theta_1 + \theta_2 + \theta_3 + \cdots}{R} \tag{5-4}$$

i.e., for series flow, we either add resistances or add driving forces.

Where there are resistance factors in parallel, presenting alternative paths of action for the driving force, let the conductances of the parallel paths be represented by

$$C_1 = \frac{1}{r_1} \qquad C_2 = \frac{1}{r_2} \qquad C_3 = \frac{1}{r_3} \qquad \cdots$$

the over-all conductance being $C = 1/R$. The fractions of the resultant (e.g., rates of heat flow) corresponding to each path are H_1, H_2, H_3, ...

The potential equation then takes the parallel form, either

$$H = \theta(C_1 + C_2 + C_3 + \cdots) \tag{5-5}$$

or

$$H_1 + H_2 + H_3 + \cdots = C\theta \tag{5-6}$$

i.e., for parallel flow we add conductances, flows, or total quantities (e.g., displacements in a static system under stress).

As in electricity, a single system may contain a combination of series and parallel processes. The potential equation then assumes a series-parallel form, e.g.,

$$H_1 + H_2 + H_3 + \cdots = \frac{\theta_1}{r_1 + r_2 + \cdots} + \frac{\theta_2}{r_3 + r_4 + \cdots}$$
$$+ \frac{\theta_3}{r_5 + r_6 + \cdots} \tag{5-7}$$

The dimensions of the driving forces and resistance or conductance factors in a physical equation depend upon the dimensions of the resultant.   The resultant may be:

1. A displacement or total quantity, i.e., a zero differential with respect to time

2. A rate or velocity, i.e., a first differential with respect to time

3. An acceleration, i.e., a second differential with respect to time

A single equation may contain all three orders of time differential, each being associated with other variables in such a manner that the terms of the equation are dimensionally homogeneous.   For example, the dynamic equation for the distance s traveled by a moving body after a time t_1 is

$$s = s_0 + vt_1 + \tfrac{1}{2}ft_1^2 \tag{5-8}$$

which can be written

$$s_0 + t_1 \frac{ds}{dt} + \frac{t_1^2}{2} \frac{d^2s}{dt^2} = ut_1 \tag{5-9}$$

where u is the average velocity over the time t_1.   This equation is of the same type as Eq. (5-6), i.e., it has the parallel form.   Successive terms contain zero, first, and second differentials with respect to time, although each term is dimensionally equivalent to a length.   In the second term on the left, the velocity v could be called the "driving force" and the time t the "conductance"; in the third term, the "driving force" would be f and the "conductance" $t_1^2/2$.   The nature of the various resistance or conductance factors in a system undergoing change is important because it fixes the *regime* of the system (Chap. 6).

Any real physical or chemical change involves more than one process, e.g., the rate of a chemical reaction depends not only on the chemical potential but also on the rates of heat and mass transfer. There will therefore be several resistance or conductance factors or alternatively several driving forces, flows, or total displacements, and it is the ratios of these factors which determine the physical "shape" of the system and which constitute the criteria of similarity.   The general method of deriving similarity criteria from the differential equations which has been described earlier exhibits these ratios as dimensionless groups.

For processes in series, the physical equations have the series form, and the dimensionless criteria are ratios of driving forces or resistances; for processes in parallel, the physical equations have the parallel form, and the dimensionless criteria are ratios of flows, conductances, or total quantities. For series-parallel processes, the compound equation [e.g., Eq. 5-7] must be resolved into its component simple equations, each of which yields one or more dimensionless criteria.

The foregoing principles will now be applied to various general types of process, viz.:

Mechanical processes

1. Elastic deformation of solids
2. Plastic deformation of solids
3. Flow of granular solids
4. Flow of fluids

Thermal processes

5. Conduction in solids
6. Conduction and convection in fluids
7. Conduction, convection, and radiation in fluids

Diffusional processes

8. Forced convection in fluids

Chemical processes

9. Homogeneous reactions
10. Heterogeneous reactions

MECHANICAL PROCESSES

Elastic Deformation of Solids

The differential equations for the three-dimensional deformation of a solid body are

$$\frac{N}{1-2\sigma}\left[\underbrace{\frac{\partial}{\partial x}\left(\frac{\partial s_x}{\partial x}+\frac{\partial s_y}{\partial y}+\frac{\partial s_z}{\partial z}\right)}_{\text{I}}\right] + N\underbrace{\left(\frac{\partial^2 s_x}{\partial x^2}+\frac{\partial^2 s_x}{\partial y^2}+\frac{\partial^2 s_x}{\partial z^2}\right)}_{\text{II}}$$

$$+ \mu\underbrace{\frac{\partial}{\partial t}\left(\frac{\partial^2 s_x}{\partial x^2}+\frac{\partial^2 s_x}{\partial y^2}+\frac{\partial^2 s_x}{\partial z^2}\right)}_{\text{III}} = \rho\underbrace{\frac{\partial^2 s_x}{\partial t^2}}_{\text{IV}} + \underbrace{\rho g\cos\alpha_x}_{\text{V}} \quad (5\text{-}10)$$

with similar equations for the y and z axes and where

N, σ, ρ, μ = modulus of rigidity, Poisson's ratio, density, viscosity of solid

s_x, s_y, s_z = displacement along x, y, z axes

t = time

g = acceleration of gravity (or equivalent body force)

α_x = angle of x axis to direction of action of body force

Each term in Eq. (5-10) represents a force parallel to the x axis acting on unit volume, viz.:

I	Elastic resistance to change of volume
II	Elastic resistance to change of shape
III	Viscous resistance to shear
IV	Inertial force
V	Gravitational or body force

The generalized dimensional form of Eq. (5-10) is

$$\underset{\text{I}}{\left[\frac{N}{(1-2\sigma)L}\right]} + \underset{\text{II}}{\left[\frac{N}{L}\right]} + \underset{\text{III}}{\left[\frac{\mu}{Lt}\right]} = \underset{\text{IV}}{\left[\frac{\rho L}{t^2}\right]} + \underset{\text{V}}{[\rho g]} \qquad (5\text{-}11)$$

where L = length, or linear dimension.

If there is an externally applied force system denoted by the force dimension F, this must be included, but expressed per unit volume like the other forces, so that (5-11) becomes

$$\underset{\text{I}}{\left[\frac{N}{(1-2\sigma)L}\right]} + \underset{\text{II}}{\left[\frac{N}{L}\right]} + \underset{\text{III}}{\left[\frac{\mu}{Lt}\right]} = \underset{\text{IV}}{\left[\frac{\rho L}{t^2}\right]} + \underset{\text{V}}{[\rho g]} + \underset{\text{VI}}{\left[\frac{F}{L^3}\right]} \qquad (5\text{-}12)$$

There are thus four independent dimensionless groups. Putting $L/t = v$ (velocity) and dividing I or II by III, IV, V and VI, respectively, the dimensionless equation corresponding to Eq. (5-12) may be written

$$\phi\left(\frac{NL}{\mu v}, \frac{N}{\rho v^2}, \frac{N}{\rho g L}, \frac{NL^2}{F}\right) = \text{const} \qquad (5\text{-}13)$$

with N replaced by $N/(1 - 2\sigma)$ if the elastic movement is purely longitudinal stretching.

For time-invariant (static) systems under stress, the inertial and viscous forces vanish, and Eq. (5-13) reduces to

$$\phi\left(\frac{N}{\rho g L}, \frac{NL^2}{F}\right) = \text{const} \qquad (5\text{-}14)$$

These results illustrate one of the advantages of starting from the fundamental differential equations rather than from the variables themselves. For the purposes of model laws, the choice of the appropriate elasticity modulus is important, and dimensional analysis alone could not have shown whether Young's modulus, the bulk modulus, or the modulus of rigidity should be used, since they all have the same dimensions.

Plastic Deformation of Solids

In the plastic deformation of solids, the applied forces are generally high compared with gravitational forces, in which case the latter can be neglected. Since the material is deformed beyond its elastic limit, Hook's law no longer describes the relation between stress and strain. In the case of a system in which only time-increasing strains occur, i.e., a nonoscillatory system, terms I and II of Eq. (5-10) can for the present purpose be replaced by a term representing stress as an undefined function ϕ of the strain and yield point, and the differential equation for plastic deformation may be written

$$\underbrace{\frac{\partial}{\partial x}\left[\phi\left(\frac{\partial s_x}{\partial x} + \frac{\partial s_y}{\partial y} + \frac{\partial s_z}{\partial z}\right)\right]}_{\text{I}} + \underbrace{\mu \frac{\partial}{\partial t}\left(\frac{\partial^2 s_x}{\partial x^2} + \frac{\partial^2 s_x}{\partial y^2} + \frac{\partial^2 s_x}{\partial z^2}\right)}_{\text{II}} = \underbrace{\rho \frac{\partial^2 s_x}{\partial t^2}}_{\text{III}}$$

$$(5\text{-}15)$$

with similar equations for the y and z axes and symbols as for Eq. (5-10).

In a solid body subject to plastic deformation, there will in general be a boundary zone on the one side of which the deformation is elastic, while on the other it is plastic. For similarity, this zone must occupy corresponding positions in prototype and model. At the boundary zone, the stress must be equal to the yield point Y of the solid material. Hence, for similarity, the scale of the first term in the generalized dimensional equation is set by Y. For the externally applied force system, an additional term must be introduced as in Eq. (5-12). The generalized dimensional equation may thus be written

$$\underbrace{\left[\frac{Y}{L}\right]}_{\text{I}} + \underbrace{\left[\frac{\mu}{Lt}\right]}_{\text{II}} = \underbrace{\left[\frac{\rho L}{t^2}\right]}_{\text{III}} + \underbrace{\left[\frac{F}{L^3}\right]}_{\text{IV}} \tag{5-16}$$

There are three independent dimensionless groups. Putting $L/t = v$ and rearranging,

$$\phi\left(\frac{Yt}{\mu}, \frac{Y}{\rho v^2}, \frac{YL^2}{F}\right) = \text{const} \tag{5-17}$$

In the static case where inertial and viscous forces disappear

$$\phi\left(\frac{YL^2}{F}\right) = \text{const} \tag{5-18}$$

Hence, the only condition for static similarity in geometrically similar solid bodies subject to plastic deformation is that the ratio of the yield stress of the material to the external forces per unit area shall be equal in prototype and model.

Movement of Granular Solids

The flow of coal, minerals, and powders down inclined chutes, the shape and density of granular beds and heaps, and the degree of segregation of material according to particle size are processes which can be studied by means of model experiments.

In this case, the forces concerned are inertia, gravity, and solid friction. The differential equation for motion along the sliding surface takes the form

$$\frac{d^2x}{dt^2} = \rho g \sin \theta - f\rho g \cos \theta \tag{5-19}$$

where f = coefficient of friction

x = linear dimension in the direction of motion over the sliding surface

θ = angle between the sliding surface and the vertical

In generalized dimension form,

$$\underset{\text{I}}{\left[\frac{\rho v^2}{L}\right]} = \underset{\text{II}}{[\rho g]} - \underset{\text{III}}{[f\rho g]} \tag{5-20}$$

Dividing I and III by II gives the dimensionless equation

$$\phi\left(\frac{v^2}{Lg}, f\right) = \text{const} \tag{5-21}$$

This means that, if the velocities of the particles are derived from falling and the heights of fall are scaled down in the same proportion as the particle size and the conformation of the surface on which they fall, then for similarity it is only necessary to have equal coefficients of friction. The densities of the falling particles in the prototype and model systems need not be equal provided that the particles are not so small that air friction becomes an important force.

Flow of Fluids

The fundamental differential equations for the isothermal flow of a Newtonian viscous fluid are the Navier-Stokes equations

$$\underset{\text{I}}{\rho\frac{\partial u}{\partial t}} + \underset{\text{II}}{\rho\left(u\frac{\partial u}{\partial x} + v\frac{\partial u}{\partial y} + w\frac{\partial u}{\partial z}\right)} = \underset{\text{III}}{\rho g \cos \alpha_x} - \underset{\text{IV}}{\frac{\partial p}{\partial x}}$$

$$+ \underset{\text{V}}{\frac{1}{3}\mu\left(\frac{\partial u}{\partial x} + \frac{\partial v}{\partial y} + \frac{\partial w}{\partial z}\right)} + \underset{\text{VI}}{\mu\left(\frac{\partial^2 u}{\partial x^2} + \frac{\partial^2 u}{\partial y^2} + \frac{\partial^2 u}{\partial z^2}\right)} \tag{5-22}$$

with corresponding equations for components parallel to the y and z axes and where

ρ, μ = fluid density, viscosity

u, v, w = velocity components parallel to the x, y, z axes, respectively

p = pressure

t = time

g = acceleration due to gravity acting at an angle α_x to the x axis

The Navier-Stokes is a dynamic equation in which each term has the dimensions of a force. Successive terms represent, respectively:

I   Force required to accelerate unit mass of fluid where the flow is unsteady
II   Transport of momentum by fluid flowing through unit cross-section area
III   Gravitational body force
IV   Static-pressure gradient
V   Viscous resistance to change of volume of the fluid (negligible for liquids)
VI   Viscous resistance to shear

The generalized dimensional equation may be written

$$\underset{\text{I}}{\left[\frac{\rho v}{t}\right]} + \underset{\text{II}}{\left[\frac{\rho v^2}{L}\right]} = \underset{\text{III}}{[\rho g]} - \underset{\text{IV}}{\left[\frac{\Delta p}{L}\right]} + \underset{\text{V VI}}{\left[\frac{\mu v}{L^2}\right]} \tag{5-23}$$

I and II are dimensionally equivalent; hence, there will be three independent dimensionless groups:

$\dfrac{\text{II}}{\text{V}}$ gives $\dfrac{\rho v L}{\mu}$ . . . . *Reynolds number*, which is the ratio of inertial to viscous forces

$\dfrac{\text{II}}{\text{III}}$ gives $\dfrac{v^2}{Lg}$ . . . . *Froude group*, which is the ratio of inertial to gravitational forces

$\dfrac{\text{IV}}{\text{II}}$ gives $\dfrac{\Delta p}{\rho v^2}$ . . . . *Pressure coefficient*, which is the ratio of pressure to inertial forces

The dimensionless equation may therefore be written

$$\phi\left(\frac{\rho v L}{\mu}, \frac{v^2}{Lg}, \frac{\Delta p}{\rho v^2}\right) = \text{const} \tag{5-24}$$

$$\frac{\Delta p}{\rho v^2} = \phi'\left(\frac{\rho v L}{\mu}, \frac{v^2}{Lg}\right) \tag{5-25}$$

This is the solution derived in Chap. 4 by the method of indices, but without the surface-tension term, since in the Navier-Stokes equations surface-tension effects are neglected. (In dynamic terms, the *Weber group* $\rho v^2 L/\sigma$ represents the ratio of inertial to surface-tension forces.) Other factors not taken into account include variations in viscosity with rate of shear, changes in viscosity and density due to the frictional heating, and compression-wave phenomona, in which elasticity plays a part. Where any of the above factors significantly influence the fluid motion, additional terms must be introduced into the equation.

Derivation of the dimensionless groups from the differential equation of motion make it clear that the Reynolds number represents the ratio of inertial to viscous forces, the Froude group the ratio of inertial to gravitational forces, and the pressure coefficient the ratio of pressure to inertial forces.

THERMAL PROCESSES

Conduction in Solids

The differential equation for conduction of heat is

$$\rho c_p \frac{\partial T}{\partial t} = k_x \frac{\partial^2 T}{\partial x^2} + k_y \frac{\partial^2 T}{\partial y^2} + k_z \frac{\partial^2 T}{\partial z^2} \tag{5-26}$$

For isotropic conductors,

$$\underset{\text{I}}{\rho c_p \frac{\partial T}{\partial t}} = k \underset{\text{II}}{\left(\frac{\partial^2 T}{\partial x^2} + \frac{\partial^2 T}{\partial y^2} + \frac{\partial^2 T}{\partial z^2} \right)} \tag{5-27}$$

where ρ, c_p = density, specific heat of solid
k = thermal conductivity
T = temperature
t = time

The terms of the equation represent, respectively:

I   Rate of increase in enthalpy per unit volume
II   Rate of conduction into unit volume

The dimensional form of the equation is

$$\left[\frac{\rho c_p T}{t} \right] = \left[\frac{kT}{L^2} \right] \tag{5-28}$$

whence the dimensionless group

$$\phi \left(\frac{\rho c_p L^2}{kt} \right) = \text{const} \tag{5-29}$$

Forced Convection in Fluids

The differential equations of heat flow by forced convection in a moving fluid are the Navier-Stokes equations [Eq. (5-22)] together with the heat-transfer equation

$$\rho c_p \left(u\,\frac{\partial T}{\partial x} + v\,\frac{\partial T}{\partial y} + w\,\frac{\partial T}{\partial z} \right) + k \left(\frac{\partial^2 T}{\partial x^2} + \frac{\partial^2 T}{\partial y^2} + \frac{\partial^2 T}{\partial z^2} \right)$$

$$= -\rho c_p \frac{\partial T}{\partial t} \quad (5\text{-}30)$$

where $\rho,\ c_p,\ k$ = density, specific heat at constant pressure, thermal conductivity of fluid

T = temperature

$u,\ v,\ w$ = fluid-velocity components parallel to $x,\ y,\ z$ axes

t = time

Successive terms in the equation represent:

I	Rate of heat loss by convection
II	Rate of heat loss by conductior
III	Rate of change of enthalpy

All based on unit volume of fluid.

Putting Eq. (5-30) into dimensional form,

$$\left[\frac{\rho c_p v T}{L} \right]_{\text{I}} + \left[\frac{kT}{L^2} \right]_{\text{II}} = - \left[\frac{\rho c_p T}{t} \right]_{\text{III}} \quad (5\text{-}31)$$

Term III represents the rate of flow of heat into or out of unit volume of fluid; it can be replaced by $[Q/L^3 T]$, where Q is the total quantity of heat transferred. If h represents a heat-transfer coefficient [e.g., in Btu/(ft^2)(hr)(°F)], Q is dimensionally equivalent to $[hL^2 Tt]$ and $[Q/L^3 t]$ to $[hT/L]$. Equation (5-31) thus becomes

$$\left[\frac{\rho c_p v T}{L} \right]_{\text{I}} + \left[\frac{kT}{L^2} \right]_{\text{II}} = - \left[\frac{hT}{L} \right]_{\text{III}} \quad (5\text{-}32)$$

Dividing I and III by II and rearranging,

$$\phi \left(\frac{\rho c_p v L}{k}, \frac{hL}{k} \right) = \text{const} \quad (5\text{-}33)$$

The group $\rho c_p v L/k$ in Eq. (5-33) represents the ratio of heat-transfer rates by conduction and bulk flow of fluid, respectively. It is known as the *Peclet group*. Equality of the Peclet group in model and prototype systems ensures that the ratios of heat flows by conduction and forced convection across corresponding surfaces shall be equal, which is one of the conditions for thermal similarity.

The other condition for thermal similarity is that the fluid-flow patterns shall be similar, i.e., that there shall be *kinematic similarity* (see Chap. 3).   The dimensionless criteria for dynamic (and hence kinematic) similarity were derived from the Navier-Stokes equations. They are the Reynolds number, representing the ratio of inertial to viscous forces, and the Froude group, representing the ratio of inertial to gravitational forces.   The full dimensionless equation for heat transfer by forced convection may therefore be written

$$\phi\left(\frac{\rho v L}{\mu}, \frac{v^2}{Lg}, \frac{\rho c_p v L}{k}, \frac{hL}{k}\right) = \text{const} \qquad (5\text{-}34)$$

The Reynolds number and Peclet group both contain vL, the product of the fluid velocity and linear dimension.   It is usually convenient that the directly controllable variables in an experimental system should appear in as few of the dimensionless group as possible.   One of the vL products can be eliminated if the Peclet group is divided by the Reynolds number.   Eq. (5-34) then becomes

$$\phi\left(\frac{\rho v L}{\mu}, \frac{v^2}{Lg}, \frac{c_p\mu}{k}, \frac{hL}{k}\right) = \text{const} \qquad (5\text{-}35)$$

or
$$\frac{hL}{k} = \phi'\left(\frac{\rho v L}{\mu}, \frac{v^2}{Lg}, \frac{c_p\mu}{k}\right) \qquad (5\text{-}36)$$

The group $c_p\mu/k$ is the *Prandtl group* and contains only physical properties of the fluid.   In model and prototype systems where both Reynolds and Prandtl groups are equal, it follows algebraically that the Peclet groups are also equal.   In closed systems with a single fluid phase, the effects of gravitational forces are negligible, and the Froude group drops out, leading to a generalized form of the standard Dittus-Boelter equation for heat transfer by forced convection:

$$\frac{hL}{k} = \phi'\left(\frac{\rho v L}{\mu}, \frac{c_p\mu}{k}\right) \qquad (5\text{-}37)$$

The Prandtl group represents the ratio of thermal diffusivity $\rho c_p/k$ to "momentum diffusivity," or kinematic viscosity, μ/ρ.   Its value for liquids varies widely owing to the wide range of liquid viscosities.   In the case of gases, the Prandtl group is relatively constant; for most gases at pressures not far from atmospheric, its value is in the region of 0.77 and independent of temperature.   The group $hL/k = h/kL\,\Delta T$ is known as the *Nusselt group*.

Radiation

The radiation of heat is a process of action at a distance and not of transfer from point to point. Hence, with respect to space coordinates, the rate equation takes an integral and not a differential form. It may be written

$$H = \frac{dQ}{dt} = \sigma e(T_1{}^4 - T_2{}^4)L^2 \tag{5-38}$$

where Q = net quantity of heat transferred
t = time
H = rate of heat transfer
T_1, T_2 = absolute temperatures of hot and cold surfaces
e = combined emissivities of hot and cold surfaces
L^2 = radiating-surface area
σ = Stefan-Boltzmann constant

Dimensionally,

$$\left[\frac{Q}{L^2t}\right] = [\sigma e T_1{}^4]\left[\frac{T_1{}^4 - T_2{}^4}{T_1{}^4}\right] \tag{5-39}$$

whence the dimensionless equation

$$\phi\left(\frac{Q}{\sigma e L^2 T^4 t}, \frac{T_1}{T_2}\right) = \text{const} \tag{5-40}$$

or

$$\phi\left(\frac{H}{\sigma e L^2 T^4}, \frac{T_1}{T_2}\right) = \text{const} \tag{5-41}$$

Radiation of heat is inevitably accompanied by conduction to and from the radiating surfaces; so the dimensionless group of Eq. (5-29) must be included:

$$\phi\left(\frac{H}{\sigma e L^2 T^4}, \frac{T_1}{T_2}, \frac{\rho c_p L^2}{kt}\right) = \text{const} \tag{5-42}$$

In a continuous-flow system such as a furnace, heat is also transported by bulk movement of fluid. The rate of bulk transport is proportional to $\rho c_p v L^2 \, \delta T$, where T is the temperature change undergone by the fluid. Provided that T_1/T_2, the ratio of corresponding pairs of temperatures, is constant, δT is proportional to T, and the ratio of total heat transferred to bulk transport of heat may be written as the dimensionless group

$$\frac{H}{v L^2 \rho c_p T}$$

Inserting this group in (Eq. (5-42), substituting $v = L/t$, and rearranging,

$$\phi\left(\underset{\text{I}}{\frac{\rho c_p v}{\sigma e T^3}}, \underset{\text{II}}{\frac{T_1}{T_2}}, \underset{\text{III}}{\frac{\rho c_p L v}{k}}, \underset{\text{IV}}{\frac{H}{kLT}}\right) = \text{const} \qquad (5\text{-}43)$$

The significance of the terms in the equation is:

I	Ratio of bulk transport to radiation
II	Absolute temperature ratio
III	Ratio of bulk transport to conduction
IV	Ratio of total heat transferred to conduction

Terms III and IV have already been met with as the Peclet and Nusselt groups, respectively. Term I is a group employed by Thring (64) which may be termed the *radiation group*.

Equation (5-43) would apply, for example, to a furnace in which the gas flow could be taken as streamline and the effects of turbulence neglected. Owing to the high kinematic viscosity of hot gases, an approximation to this condition is not uncommon (see Chap. 7).

DIFFUSIONAL PROCESSES

Forced Convection in Fluids

The equations for mass transfer are analogous to those for heat transfer, concentration taking the place of enthalpy and diffusivity replacing conductivity. The differential equations for mass transfer with forced convection are the Navier-Stokes equations [Eq. (5-22)] together with the mass-transfer equation

$$\underset{\text{I}}{\left(u\frac{\partial a}{\partial x} + v\frac{\partial a}{\partial y} + w\frac{\partial a}{\partial z}\right)} + \underset{\text{II}}{D\left(\frac{\partial^2 a}{\partial x^2} + \frac{\partial^2 a}{\partial y^2} + \frac{\partial^2 a}{\partial z^2}\right)} = \underset{\text{III}}{-\frac{\partial a}{\partial t}} \qquad (5\text{-}44)$$

where a = concentration of diffusing substance per unit volume
$\quad\ D$ = diffusion coefficient of diffusing substance
and other symbols are as for Eq. (5-30).

Equation (5-44) is in all respects analogous to Eq. (5-30). Successive terms represent:

I	Rate of mass transfer by convection
II	Rate of mass transfer by diffusion
III	Rate of change of concentration

In dimensional form,

$$\left[\frac{va}{L}\right] + \left[\frac{aD}{L^2}\right] = -\left[\frac{a}{t}\right] \qquad (5\text{-}45)$$

Terms I and III are dimensionally equivalent; hence, there is one dimensionless group which may be written, for the steady state (time-invariant),

$$\phi\left(\frac{vL}{D}\right) = \text{const} \tag{5-46}$$

or, for the unsteady state,

$$\phi\left(\frac{L^2}{Dt}\right) = \text{const} \tag{5-47}$$

Combining Eq. (5-44) with the dimensionless equation for the steady state derived from the Navier-Stokes equations [Eq. (5-22)],

$$\phi\left(\frac{\rho vL}{\mu}, \frac{v^2}{Lg}, \frac{\Delta p}{\rho v^2}, \frac{vL}{D}\right) = \text{const} \tag{5-48}$$

The expression vL occurs in both the first and the last group of Eq. (5-48). As was done in the case of Eqs. (5-34) and (5-36), it is convenient to divide the final group by the Reynolds number, yielding a group which contains only physical properties of the fluid. Equation (5-48) then becomes

$$\phi\left(\frac{\rho vL}{\mu}, \frac{v^2}{Lg}, \frac{\Delta p}{\rho v^2}, \frac{\mu}{\rho D}\right) = \text{const} \tag{5-49}$$

The group $\mu/\rho D$ is known as the *Schmidt group*.

CHEMICAL PROCESSES

Homogeneous Reactions

Chemical reactions take place in fluid media and are accompanied by the evolution or absorption of heat. The reaction rate is strongly influenced by the temperature. It is also affected by the point concentration of reactants and reaction products, and hence by fluid-flow patterns and rates of mass transfer. The complete mathematical specification of a homogeneous chemical process therefore requires the Navier-Stokes equations governing fluid flow, an equation connecting heat-transfer rate with rate of heat evolution, and a further equation connecting mass-transfer rate with rate of consumption of reactants or formation of products. The equations which relate chemical-reaction rate to temperature and concentration of reactants should also be taken into account.

An equation connecting heat transfer by conduction and convection with rate of heat evolution is derived from Eq. (5-29),

$$\rho c_p \left(u \frac{\partial T}{\partial x} + v \frac{\partial T}{\partial y} + w \frac{\partial T}{\partial z} \right) + k \left(\frac{\partial^2 T}{\partial x^2} + \frac{\partial^2 T}{\partial y^2} + \frac{\partial^2 T}{\partial z^2} \right) + \rho c_p \frac{\partial T}{\partial t} = qU \tag{5-50}$$

where U = chemical-reaction rate expressed as mass of product formed per unit volume and time

q = heat of reaction per unit mass of product

Other symbols are as for Eq. (5-30). Dimensionally,

$$\left[\frac{\rho c_p v T}{L} \right] + \left[\frac{kT}{L^2} \right] + \left[\frac{\rho c_p T}{t} \right] = [qU] \tag{5-51}$$

whence the dimensionless equation

$$\phi \left(\frac{\rho c_p v L}{k}, \frac{qUL}{\rho c_p v T} \right) = \text{const} \tag{5-52}$$

The mass-transfer equation is similarly derived from Eq. (5-44),

$$\left(u \frac{\partial a}{\partial x} + v \frac{\partial a}{\partial y} + w \frac{\partial a}{\partial z} \right) + D \left(\frac{\partial^2 a}{\partial x^2} + \frac{\partial^2 a}{\partial y^2} + \frac{\partial^2 a}{\partial z^2} \right) + \frac{\partial a}{\partial t} = U \tag{5-53}$$

whence the dimensionless equation

$$\phi \left(\frac{vL}{D}, \frac{UL}{av} \right) = \text{const} \tag{5-54}$$

Heat transfer by radiation cannot in general be neglected, and the dimensionless groups of Eq. (5-40) must also be taken into account. Thus, the full dimensionless equation for a homogeneous chemical reaction taking place in a moving-fluid medium is as follows:

$$\phi \left(\underset{\text{I}}{\frac{\rho v L}{\mu}}, \underset{\text{II}}{\frac{v^2}{2g}}, \underset{\text{III}}{\frac{\Delta P}{\rho v^2}}, \underset{\text{IV}}{\frac{\rho c_p v L}{k}}, \underset{\text{V}}{\frac{qUL}{\rho c_p v T}}, \underset{\text{VI}}{\frac{vL}{D}}, \underset{\text{VII}}{\frac{UL}{av}}, \underset{\text{VIII}}{\frac{Q}{\sigma e L^2 T^4}}, \underset{\text{IX}}{\frac{T_1}{T_2}} \right) = \text{const} \tag{5-55}$$

If the chemical reaction is the only source of heat, Q in term VIII can be replaced by qUL^3. Dividing terms III and VI by I converts them into the Prandtl and Schmidt groups, respectively. To eliminate U from all groups but one, divide terms V and VIII by VII. Gravitational and frictional effects in chemically reacting systems are generally negligible, in which case terms II and III can be omitted. With these changes and some rearrangement, Eq. (5-55) becomes

$$\underset{\text{I}}{\frac{UL}{av}} = \phi \left(\underset{\text{II}}{\frac{\rho v L}{\mu}}, \underset{\text{III}}{\frac{c_p \mu}{k}}, \underset{\text{IV}}{\frac{\mu}{\rho D}}, \underset{\text{V}}{\frac{qa}{\rho c_p T}}, \underset{\text{VI}}{\frac{qav}{\sigma e T^4}}, \underset{\text{VII}}{\frac{T_1}{T_2}} \right) \tag{5-56}$$

The significance of the terms in Eq. (5-56) is:

I Ratio of product concentration to reactant concentration after a given time or length of travel (e.g., in a tubular reactor)

II Reynolds number, representing the ratio of inertial to viscous forces and governing the flow pattern and hence the statistical distribution of residence times among the reactant molecules

III Prandtl number, representing the ratio of heat transferred by bulk transport to heat transferred by conduction

IV Schmidt number, representing the ratio of material transferred by bulk transport to material transferred by molecular diffusion

V Ratio of potential-chemical-heat content to sensible-heat content per unit volume

VI Ratio of potential chemical heat transferred by bulk transport to heat radiated

VII Ratio of emission rates of hot and cold radiating surfaces

Where the effects of radiation are negligible, terms VI and VII can be omitted and the resulting equation is dimensionally equivalent to that derived by Damköhler for a continuous reactor (7),

$$\frac{UL}{av} = \phi\left(\frac{\rho v L}{\mu}, \frac{c_p \mu}{k}, \frac{\mu}{\rho D}, \frac{qa}{\rho c_p T}\right) \tag{5-57}$$

Alternatively, if heat transfer by conduction is negligible compared with radiation, as may be the case in combustion and other high-temperature gas reactions, terms III and V can be omitted and the dimensionless equation becomes

$$\frac{UL}{av} = \phi\left(\frac{\rho v L}{\mu}, \frac{\mu}{\rho D}, \frac{qav}{\sigma e_T T^4}, \frac{T_1}{T_2}\right) \tag{5-58}$$

Reaction Rate

The reaction rate U is not itself a directly controllable variable but depends on the controllable variables' temperature and concentration. The general equation for the rate of a homogeneous chemical reaction is

$$U = K_n F(a_1 a_2 \cdots a_n) \tag{5-59}$$

where K_n = velocity constant of nth-order reaction
$a_1 a_2 \cdots a_n$ = molecular concentration of reactants
F = dimensionless kinetic factor derived from activity coefficients of reactants and products

The dimensions of K_n vary with the order of the reaction so that the product $K_n(a_1 a_2 \cdots a_n)$ always has the dimensions $mL^{-3}t^{-1}$ (see Appendix 2). The value of K_n is strongly influenced by the reaction temperature in accordance with the modified Arrhenius equation (136),

$$K_n = A T^{1/2} \epsilon^{-E/RT} \tag{5-60}$$

where A = a dimensional constant

$\quad \epsilon$ = 2.718

$\quad E$ = energy of activation

$\quad R$ = gas constant

$\quad T$ = absolute temperature

Combining Eqs. (5-59) and (5-60) and taking the concentrations $a_2 \cdots a_n$ as being each proportional to $a_1 = a$, the group UL/av in Eqs. (5-57) and (5-58) is replaced by two groups,

$$\frac{Ka^{2n-2}L^2T}{v^2} \qquad \frac{E}{RT}$$

or, for the unsteady state (e.g., in a batch process),

$$Ka^{2n-2}t^2T \qquad \frac{E}{RT}$$

where $K = A^2F^2$ (K is thus a dimensional constant having the dimensions $m^{2n-2}L^{6-6n}t^{-2}T^{-1}$).

The introduction of the Arrhenius relation does not increase the number of dimensionless groups required to specify complete chemical similarity, and the "Arrhenius group" E/RT may either replace one of the other temperature groups or be combined with it. For example, the temperature may be eliminated from term V in Eq. (5-56),

$$\frac{qa}{\rho c_p T} \div \frac{E}{RT} = \frac{qaR}{Ec_p}$$

Equation (5-56) becomes

$$\underset{\text{I}}{\frac{Ka^{2n-2}L^2T}{v^2}} = \phi \left(\underset{\text{II}}{\frac{\rho vL}{\mu}}, \underset{\text{III}}{\frac{c_p\mu}{k}}, \underset{\text{IV}}{\frac{\mu}{\rho D}}, \underset{\text{V}}{\frac{qaR}{E\rho c_p}}, \underset{\text{VI}}{\frac{qa}{\sigma eT^4}}, \underset{\text{VII}}{\frac{T_1}{T_2}} \right) \qquad (5\text{-}61)$$

For batch reaction, group I is replaced by $Ka^{2n-2}t^2T$, where t is the reaction time.

Equation (5-61) applies where there is only one irreversible reaction or, in the case of simultaneous reactions, where they are all of the same order. If several reactions of different orders occur simultaneously, then group I would be replaced by several groups, one for each order of reaction. This, together with the fact that the order of some or all of the constituent reactions may be unknown, at present limits the practical utility of equations such as (5-61).

Heterogeneous Reactions

The rate of a heterogeneous chemical reaction depends upon the interfacial area between the phases. Putting

U' = heterogeneous reaction rate expressed as mass of product per unit interfacial area in unit time

s = specific surface or interface per unit volume

group I in Eq. (5-56) becomes

$$\frac{U'sL}{av}$$

or, for geometrically similar systems ($s \propto 1/L$),

$$\frac{U'}{av}$$

The velocity of a heterogeneous chemical reaction is a complex function of reactant concentration and is affected by such factors as the extent to which reactants and products are absorbed at the interface and any specific catalytic activity that the interface may possess. The relation may be represented by a generalized equation

$$U' = \Phi[K_n' F\alpha(a_1 a_2 \cdots a_n)] \tag{5-62}$$

where Φ = an unknown function

K_n' = velocity constant for a heterogeneous reaction of nth order

F = kinetic factor as for homogeneous reactions

α = a dimensionless factor proportional to catalytic activity of interface

The dimensions of K_n' vary with the order of the reaction so that the product $K_n'(a_1 a_2 \cdots a_n)$ always has the dimensions $mL^{-2}t^{-1}$.

The heterogeneous reaction rate U' conforms to a modified Arrhenius equation in which E is replaced by the apparent energy of activation E', which includes effects due to heats of absorption. Group I in Eq. (5-61) thus becomes for heterogeneous reactions

$$\frac{K'\alpha s a^{2n-2}L^2 T}{v^2}$$

for the steady state or $K'\alpha s a^{2n-2}t^2 T$ for the unsteady state.

In this and the preceding chapters, a number of dimensionless groups have appeared, many of them bearing names. Appendix 3 gives a list of the principal dimensionless groups that are met with in model theory.

SYMBOLS IN CHAPTER 5

A = dimensional constant in modified Arrhenius equation

a = concentration, mass or moles per unit volume

a = specific surface per unit volume

c_p = specific heat at constant pressure

D = diffusion coefficient
E = activation energy
E = electromotive force
E' = apparent energy of activation (heterogeneous reactions)
e = emissivity
F = force
F = dimensionless kinetic factor
f = coefficient of solid friction
g = acceleration of gravity
H = heat flux (rate of flow)
h = heat-transfer coefficient
I = electrical current
K = homogeneous-reaction velocity constant
K' = heterogeneous-reaction velocity constant
K, K' = dimensional constant for homogeneous, heterogeneous reactions
k = thermal conductivity
L = length, linear dimension
N = modulus of rigidity
p = pressure
Q = quantity of heat
q = heat of reaction per unit mass of product
R = gas constant
R = thermal resistance (total)
R_E = electrical resistance
r = thermal resistance (individual)
s = displacement, distance traveled
T = temperature
t = time
U = chemical-reaction rate as mass of product formed per unit volume and time
u = average velocity
u, v, w = velocity components along x, y, z axes
v = velocity
x, y, z = linear coordinates at right angles
Y = yield point
α = angle
α = dimensionless catalyst activity factor
Δ = difference
δ = increment (δT)
θ = angle
θ = temperature difference ($= \Delta T$)
μ = viscosity
ρ = density
σ = Poisson's ratio
σ = Stefan-Boltzmann constant
σ = surface or interfacial tension
Φ, ϕ = function

CHAPTER 6

THE REGIME CONCEPT

In Chap. 3, physical systems were classified as static, dynamic, thermal, chemical, etc. Any real system comes under more than one of these heads according to the particular aspect of its behavior that is being considered. For example, a tubular condenser constitutes a static system with respect to tube and shell stresses, a dynamic system with respect to flow pattern and pressure drop, a thermal system with respect to heat transfer, and a chemical system with respect to scaling and corrosion. The similarity criteria will vary according to which of these effects is being studied.

In Chap. 5, it was pointed out that dimensionless similarity criteria are ratios of physical quantities which are functions of the various forces or resistances which control the reaction rate. Where there are several controlling factors of different kinds, there will be several dimensionless criteria. For example, resistance to the motion of a fluid may be due to viscous drag, gravitational forces, or surface tension, the corresponding criteria being the Reynolds, Froude, and Weber groups, respectively. For homologous systems of different absolute magnitudes, these three criteria are mutually incompatible, since each one requires the fluid velocity to vary as a different function of the linear dimensions, viz.:

For equal Reynolds numbers $v \propto L^{-1}$
For equal Froude numbers $v \propto L^{\frac{1}{2}}$
For equal Weber numbers $v \propto L^{-\frac{1}{2}}$

By using fluids with widely different physical properties in the two systems and selecting the scale ratio accordingly, it is possible, within limits, to satisfy any two of these criteria simultaneously, but seldom, if ever, all three.* Hence, in scaling up a complex physical or chemical

---

* For example, in an unbaffled paddle mixer, the flow pattern depends partly on the Reynolds number and partly on the Froude group. Both criteria can be made equal in geometrically similar mixers of different sizes provided that liquids

process, it is advantageous to choose conditions such that the rate of the whole process depends predominantly upon one dimensionless criterion.   In a purely physical model experiment for which a non-homologous system may be employed, two dimensionless criteria may enter.

The term *regime* is used in this book to distinguish the rate-determining process in a system in which several other processes in series or parallel may also be occurring, or, in other words, the particular force, flow, or resistance factor which controls the over-all rate of change. In a static system, the regime distinguishes the factors governing the total displacement.

An example of the regime concept applied to a chemical process is taken from Laupichler's study of the catalytic water-gas reaction (21). The reaction rate was found to be inversely proportional to a total reaction resistance R, defined by the equation

$$R = \frac{1}{kC_m} + \frac{\delta}{D} \tag{6-1}$$

where k = velocity constant of water-gas reaction

C_m = mean concentration of water vapor

δ = thickness of streamline gas film on catalyst surface

D = diffusion coefficient of carbon monoxide through gas film

The first term on the right was called by Laupichler the *conversion resistance* and the second term the *diffusion resistance*, and these are convenient terms for the characterization of chemical reactions in general.   In the catalytic water-gas reaction, the chemical-reaction

---

of different viscosities can be used in the two systems.

For equal Froude groups:

$$\frac{v}{v'} = \left(\frac{L}{L'}\right)^{1/2}$$

For equal Reynolds numbers:

$$\frac{v}{v'} = \frac{L'}{L}\frac{\nu}{\nu'}$$

where ν = kinematic viscosity of liquid.

Equating the two conditions,

$$\left(\frac{L}{L'}\right)^{1/2} = \frac{L'}{L}\frac{\nu}{\nu'} \quad \text{or} \quad \frac{\nu}{\nu'} = \left(\frac{L}{L'}\right)^{3/2}$$

The experimental procedure is therefore to choose two liquids having very different kinematic viscosities, make the linear scale ratio equal to the $\frac{3}{2}$ power of the kinematic viscosity ratio, employ the more viscous liquid in the larger mixer, and fix corresponding speeds to give equal Froude groups   Under these conditions, the Reynolds numbers will also be equal.

velocity was found to be slow compared with the rates of diffusion of the reactants; the over-all reaction rate was therefore controlled by the conversion resistance, the diffusion resistance being negligible. In other heterogeneous reactions such as the absorption of a gaseous ammonia in sulfuric acid, the conversion resistance is low, and the over-all rate is controlled by the diffusion resistance.

Conversion resistance is an inverse function of the chemical-reaction velocity; hence, when this factor determines the over-all reaction rate, the system is said to be subject to a *chemical regime*. If diffusion resistance is the controlling factor, the reaction rate will depend upon the fluid dynamics of the system, which is then subject to a *dynamic regime*. Where there is a chemical regime, the scale-up relation calls for chemical similarity; where there is a dynamic regime, for dynamic similarity.

The rate-determining process in a system is to be distinguished from the main process taking place, or that which it is the object of the operation to effect. Thus, chemical reactions may be subject to a dynamic regime where the system is heterogeneous and the chemical-reaction velocity is high. Similarly, heat transfer is subject to a dynamic regime where forced convection is the rate-determining process; where radiation or natural convection controls the rate, the regime is thermal. In a given system, the regime depends upon the relative magnitudes of the various reaction resistances, and these will vary with the conditions of operation. In a chemical reaction that is subject to a dynamic regime, increased agitation may eventually diminish the diffusion resistance to a point at which the conversion resistance predominates and the regime becomes chemical. Between the two lies an intermediate region of *mixed regime* in which both conversion and diffusion resistances are appreciable.

Corresponding to each main class of regime (static, dynamic, thermal, or chemical), there are several possible variants according to the nature of the driving force or resistance factor which controls the resultant (whether it be a total quantity or a rate of change). For example, a fluid dynamic regime may be controlled by the ratio of inertial forces to viscosity, gravity, or surface tension, and the viscosity-controlled regime may be either streamline or turbulent.

For the reliable scaling up or down of a complex physical or chemical process, two conditions are necessary:

1. The regime should be relatively "pure"; i.e., the reaction rate should depend chiefly upon a single dimensionless group.

2. The regime should be of the same type on both the small and the large scale.

The first condition has already been discussed.   The second requires that, in planning a series of pilot-plant or model experiments, corresponding conditions in the large-scale prototype must be constantly borne in mind.   The danger of a change in regime occurs chiefly when it is necessary to extrapolate the similarity relation (see Chap. 8).

The prevailing regime in a given system is often seen by inspection. Thus, when heat is being transferred to a liquid by natural convection alone, it is obvious that there is a thermal regime.   The only process variable affecting the rate of heat transfer is the temperature difference.   If the prevailing regime is not self-evident, then it must be determined, either theoretically or empirically.   The theoretical method consists in examining the rate equation and calculating the order of magnitude of the separate resistance factors, driving-force components, or flows which combine to determine the over-all reaction rate.   For example, in the combustion of an atomized liquid fuel, it can be calculated that under normal conditions the times required for the droplets to evaporate and then to ignite are both less than one-tenth of the time needed for mixing with the combustion air.   The whole process is therefore controlled by the diffusion resistance and is subject to a dynamic regime.

The empirical method of determining the prevailing regime consists in observing experimentally the effect of certain variables on the over-all reaction rate.   It is not then necessary to know the rate equation of the system.   This method is particularly applicable where both chemical and diffusional resistances are involved.   For such a system, the design of plant and the choice of operating conditions will depend largely on whether a chemical or a dynamic regime prevails, and the method of determining this is to observe the effect of (1) temperature change and (2) degree of agitation upon the over-all reaction rate. These two variables are of such general importance in chemical-engineering operations and processes that their effects will be discussed in some detail.

Effect of Temperature

An increase in temperature has a tendency to increase the rate of both chemical and physical reactions by diminishing the resistance factor in the generalized rate equation.   In the case of balanced chemical reactions, a temperature rise may also shift the equilibrium point in a direction which reduces the driving force of the reaction, i.e., the difference between actual and equilibrium concentrations or activities.   In exceptional cases, the reduced driving force may more than counterbalance the diminished reaction resistance, with the net result that the reaction rate displays a negative temperature coefficient. An example is the atmospheric oxidation of nitric oxide, a homogeneous reaction whose velocity is markedly diminished by increase in temperature. Generally, however, the principal effect of an unfavorable equilibrium shift is to reduce the yield of product. In such cases, a reaction temperature must be chosen which strikes an optimum balance between high reaction rate and high yield. The effect of a small temperature rise on the driving force of a chemical reaction is usually negligible compared with its effect on the reaction resistance or its reciprocal, the velocity constant.

Physical reactions such as heat or mass transfer also tend to be accelerated by a rise in temperature, especially when they take place in the liquid phase. The principal factor is the reduction in viscosity with rise in temperature; in the case of mass-transfer processes, there is also an increase in diffusivity or diffusion coefficient. For comparison with chemical-reaction velocity and diffusivity, one should consider the fluidity $1/\mu$, the reciprocal of viscosity.

Chemical-reaction velocities and the fluidity and diffusivity of liquids all conform approximately to equations of an exponential type relating them to the absolute temperature T. In the case of chemical reactions, the equation is known as the Arrhenius equation. The three equations are

$$k = A\epsilon^{-E/RT}$$

$$\frac{1}{\mu} = B\epsilon^{-E_v/RT}$$

$$D = C\epsilon^{-E_d/RT}$$

where k = chemical-reaction velocity constant
$\quad 1/\mu$ = fluidity
$\quad\quad D$ = diffusivity
$\quad\quad E$ = activation energy for chemical reaction
$\quad\quad E_v$ = activation energy for viscous flow
$\quad\quad E_d$ = activation energy for diffusion
A, B, C = constants

The above equations are approximate because neither the activation energies nor the "constants" A, B, and C are entirely independent of temperature. According to Glasstone, Laidler, and Eyring (136), $A = aT^{1/2}$, $B = bT^{3/2}$, $C = cT^{1/2}$, where a, b and c are true constants. Nevertheless, A, B, and C may be assumed constant over small temperature ranges.

The temperature coefficient of a reaction rate is commonly expressed as the relative increase in rate which is caused by a temperature rise of $10°C$. This is a convenient parameter, but for a given reaction its numerical value diminishes with rise in temperature. This can be seen by evaluating the $10°$ temperature coefficient in terms of the Arrhenius equation

$$\log \frac{k_{T+10}}{k_T} = \frac{E}{R}\left(\frac{1}{T} - \frac{1}{T + 10}\right)$$
$$= \frac{10E}{RT_m{}^2} \tag{6-2}$$

where $T_m = \sqrt{T(T + 10)}$.

A similar equation gives the $10°$ temperature coefficient of fluidity or diffusivity in liquids when E_v or E_d is substituted for E. Hence, for purposes of rough comparison, the temperature coefficient of k, $1/\mu$, or D experimentally determined at any temperature T can be corrected to a standard temperature, say, $15°C$ ($= 288°K$), by multiplying log (coefficient) by $T^2/288^2$.

At temperatures in the region of atmospheric, the numerical value of the $10°$ temperature coefficient for most chemical-reaction velocities lies between 2 and 4. The value for liquid-phase physical reactions is much less than this, for two reasons: first, activation energies for viscous flow or diffusion are smaller than activation energies for chemical reaction; second, both fluidity (or viscosity) and diffusivity (diffusion coefficient) enter the reaction-rate equations as factors in dimensionless groups raised to powers which are less than unity and may even be zero.

The temperature coefficients of mass-transfer rates in the liquid phase vary with the absolute value of the diffusion coefficient. In general, solutes possessing high diffusion coefficients have low temperature coefficients (136). For most mass-transfer processes occurring in aqueous solutions at or near atmospheric temperatures the $10°$ temperature coefficient is about 1.24.

In liquid-phase heat-transfer processes, the quantity corresponding to diffusivity is thermal conductivity, and the thermal conductivities of most liquids decrease with temperature rise, whereas diffusivities increase. Hence, the $10°$ temperature coefficients of liquid-phase convectional heat-transfer rates tend to be lower than for mass-transfer processes, usually below 1.10.

In the gaseous phase, viscosity and diffusivity no longer conform to an exponential type of equation—although chemical reactions in the gaseous phase continue to do so. The viscosity of a gas is approximately independent of pressure and proportional to the square root of the absolute temperature, i.e.,

$$\mu = B'T^{\frac{1}{2}}$$

Diffusion coefficients in gases vary inversely with pressure and with temperature according to the equation

$$D = C'T^x$$

where x is between $\frac{3}{2}$ and 2, depending on the gas.

The thermal conductivity of a gas (k) increases with temperature in a manner similar to viscosity, with which it is connected by the Maxwell equation

$$k = a\mu c_v$$

where c_v is the specific heat at constant volume and a is a constant.

The Prandtl group $c_p\mu/k$, which figures in the dimensionless equation for convectional heat transfer, is approximately independent of temperature for gases.

The net result is that the 10° temperature coefficient of heat- and mass-transfer rates in the gaseous phase is very low, of the order of 1.01 at ordinary temperatures.

Effect of Agitation

The general effect of agitation is to increase the degree of turbulence in a fluid medium, reduce the thickness of streamline boundary films, and so diminish the resistance to processes of heat or mass transfer by convection.

The generalized dimensionless rate equation for heat transfer under a viscosity-controlled dynamic regime is

$$\frac{hL}{k} = \left(\frac{\rho vL}{\mu}\right)^x \left(\frac{c_p\mu}{k}\right)^p \tag{6-3}$$

where h = film coefficient of heat transfer
L = linear dimension
v = fluid velocity, or rate of stirring
k = thermal conductivity of fluid

For mass transfer, the corresponding equation is

$$\frac{KL}{D} = \left(\frac{\rho vL}{\mu}\right)^x \left(\frac{\mu}{\rho D}\right)^q \tag{6-4}$$

where K = mass-transfer coefficient and D = diffusion coefficient.

Comparing the same system at different values of v, we have, for both heat and mass transfer,

$$\frac{h_2}{h_1} = \frac{K_2}{K_1} = \left(\frac{v_2}{v_1}\right)^x \tag{6-5}$$

For streamline flow, the value of x is zero. The laminar movement of the fluid does not contribute to the transfer of heat or matter across the streamlines, and conduction or diffusion proceed as in a stationary fluid. For turbulent flow with fixed interface, i.e., in a solid-liquid system, the value of x varies from 0.6 to 0.8 according to the geometry of the system. For turbulent flow with free interface, i.e., in a liquid-liquid or liquid-vapor system, agitation has two separate effects: it diminishes the transfer resistance at the interface and also increases the interfacial area by more thoroughly dispersing one phase in the other. In the case of immiscible liquids, the combined effects produce values of x varying from 3.8 to 5.0 according to the geometry of the system (195). Hence, it is a general characteristic of fluid processes subject to a dynamic regime that in the turbulent region the reaction rate varies as some power of the fluid velocity or rate of stirring.

Where there is a chemical regime in either a homogeneous or a heterogeneous system, the reaction rate does not depend on the rate of mass transfer and is therefore independent of the fluid velocity. Evidently, as mass-transfer coefficients and (in the case of free-interface heterogeneous systems) interfacial area decrease, a point will be reached at which the regime changes from chemical to dynamic and the reaction rate then becomes dependent on the rate of agitation.

The Reynolds-number exponent x in the dimensionless rate equation is a useful indication of the prevailing fluid regime. It will be referred to as the *Reynolds index*, defined as that power of the Reynolds number according to which a reaction rate varies at constant temperature and pressure.

The empirical method of discovering the prevailing regime where both chemical and dynamic processes take place depends upon the experimental determination of the two parameters mentioned above, the 10° temperature coefficient (corrected where necessary to 15°C), and the Reynolds index. The conclusions to be drawn from these parameters may be summarized in the following general rules:

1. *A 10° temperature coefficient greater than 2 characterizes a chemical regime; a coefficient below 1.5 characterizes a dynamic regime.*

2. *A Reynolds index approximating to zero characterizes a chemical regime or a streamline dynamic regime; an index between 0.5 and 0.8 characterizes a turbulent dynamic regime with fixed interface; an index between 3 and 5 characterizes a dynamic regime with free interface, i.e.,*

a two-phase liquid-liquid or gas-liquid system. (The similarity criterion here is likely to be the Weber group rather than the Reynolds number).

3. *Where the 10° temperature coefficient is greater than 1.5 and the Reynolds index is below 0.5, a mixed chemical-dynamic regime is indicated.*

A table showing typical values of the Reynolds index for various cases of heat and mass transfer is given in Chap. 8 (Table 8-1).

Mixed Regime

Where the 10° temperature coefficient of a reaction is below 2.0 and the Reynolds index is substantially greater than zero but less than 0.6, the indications are that both conversion and diffusion resistance have a considerable influence on the over-all reaction velocity. Or again, under a purely dynamic regime, it may be found that both gravity and viscosity significantly influence the rate of change; i.e., both Froude and Reynolds numbers have to be taken into account. These are instances of a *mixed regime.*

In general, a mixed regime exists when there are two or more reaction resistances which significantly influence the reaction rate and which conform to different linear scale relations. Therefore, if the process is scaled up with respect to one class of resistance, there is no similarity with respect to the other. This type of problem is constantly met with, and it is necessary to have recourse to various empirical methods of correcting for one class of resistance while scaling up with respect to the other. A mixed regime marks a danger point in any new process, since it is not always possible to find a reliable basis for predicting large-scale results from small-scale experiments.

It is sometimes feasible to lift a reaction out of the region of mixed regime by changing the operating conditions so that one or other class of resistance becomes negligible. Since chemical reactions have higher temperature coefficients than diffusional phenomena, increasing the temperature tends to convert a chemical regime, first into a mixed, and finally into a viscosity-controlled dynamic regime. Similarly, where there is a heterogeneous chemical regime, reducing the degree of agitation increases the diffusion resistance so that the regime tends to become first mixed, then wholly dynamic. In each case, a change of temperature or agitation, respectively, in either direction, if it could be carried far enough, would take the reaction out of the region of mixed regime. Or again, where there is mixed viscosity and gravity control, a change in the geometry of the system may eliminate the gravitational effect, as when baffles are introduced into a paddle mixer.

Where it is impossible or impracticable to escape from a mixed regime by modifying the operating conditions, there are various devices by which the inherent difficulties may be at least partially overcome. These are as follows:

1. It is sometimes possible to calculate one of two incompatible resistance factors and carry out model experiments to determine the other. The classic example is the tank testing of model ship hulls, where skin friction is calculated and deducted from the total drag, the difference being the incalculable resistance due to wave formation. Corresponding speeds can then be fixed in relation to wave formation only, i.e., assuming a gravity-controlled dynamic regime.

2. A geometrical distortion of the model can compensate for a mixed regime in some cases. For example, consider a bare tube through which a gaseous mixture is flowing and reacting chemically, heat being removed by radiation and natural convection from the outside of the tube. The over-all reaction rate will be influenced both by the chemical composition and by the rate of heat transfer; i.e., the regime is partly chemical and partly thermal. The controlling factor in the over-all rate of heat transfer is assumed to be the rate of dissipation from the outside of the tube. The reaction temperature is to be the same as both the small and the large scale, whence the heat loss per unit area of external surface must be the same. A scale model will not meet the required conditions because thermal similarity requires an equal heat flux per unit of surface area, whereas the model reaction tube will possess r times as much surface per Btu evolved per hour as the prototype. Hence, heat losses in the model will be excessive, and the reaction temperature will not be maintained.

In this case, similarity can be achieved by distorting the model so that the surface area per unit volume is the same as in the prototype. Neglecting the end effects, this requires that the diameter of both tubes shall be the same; i.e., if the prototype is a long tube, the model will be a squat vessel of the same diameter, the throughput of reactants being adjusted so as to give the same time of residence as in the prototype. Heat loss from the ends of the model vessel should be prevented by insulation. (End effects can be neglected only over a moderate range of scale ratios. Where the scale reduction is large, this principle would tend to give the model a pancake form in which end effects would actually predominate.)

Under the above circumstances, the internal coefficient of heat transfer will be lower in the model than in the prototype because of the lower fluid velocity. This difference is assumed to have a negligible influence on the over-all coefficient, but similarity would be improved by calculating approximately the internal film coefficients in both vessels and allowing sufficient extra surface in the model to compensate for the slightly decreased over-all coefficient. In other words, the second-order discrepancy due to the internal film coefficients is corrected by method 1.

3. A third method of dealing with a mixed regime is to modify one of the controlling rates by some artificial arrangement in the model. In the above example, a scale-model reaction tube could have been used if the external surface of the model had been insulated all over to reduce the rate of heat loss per unit volume of reaction space to the same value as in the uninsulated prototype. This is dealt with more fully under Boundary Effects, Chap. 9.

SYMBOLS IN CHAPTER 6

A, B, C = nominal constants
a, b, c = true constants
B', C' = nominal constants
C_m = mean concentration
c_p = specific heat at constant pressure
c_v = specific heat at constant volume
D = diffusion coefficient
E = activation energy for chemical reaction
E_d = activation energy for diffusion
E_v = activation energy for viscous flow
h = heat-transfer coefficient
K = mass-transfer coefficient
k = thermal conductivity
k = chemical-reaction velocity constant
L = linear dimension
p, q = exponents
R = gas constant
R = reaction resistance
r = linear scale ratio
T = absolute temperature
v = velocity
x = exponent on T
x = Reynolds index
δ = film thickness
ϵ = 2.718
μ = viscosity
ν = kinematic viscosity
ρ = density

CHAPTER 7

SIMILARITY CRITERIA AND SCALE EQUATIONS

This chapter brings together the similarity criteria for the principal types of regime that are met with in chemical engineering. For convenience, each criterion is followed by several scale equations giving the ratios between corresponding quantities where similarity exists. Ratios of quantities are indicated by the use of boldface type: thus **v** denotes a ratio of velocities $= v'/v$; $\boldsymbol{\varrho}$ denotes a ratio of densities $= \rho'/\rho$; the numerator is always the quantity pertaining to the large-scale system.

In homologous systems, corresponding quantities are related in terms of the scale ratio **L** only. Homologous systems are defined as systems in which:

1. The shapes of corresponding solid members or of the solid surfaces enclosing fluid masses are geometrically similar.

2. Chemical compositions and physical properties at corresponding points, in so far as they affect the process being studied, are identical.

Where the prototype apparatus has a multiple or grid structure (multitubular, granular, etc.), it is usually preferable for the small-scale unit to be an element or model element rather than a scale model (see Chap. 3), provided that wall effects are either negligible or capable of being independently regulated (Chap. 9). An element is the only type of scaled-down apparatus that can achieve similarity under a mixed regime.

Static Regime

The static regime is concerned with the deformation of solid bodies and structures under stress, a phenomenon that has often been investigated by means of experiments with scale models. The requirement for similarity is that geometrically similar bodies shall suffer geometrically similar deformations, i.e., that strains at corresponding points shall be equal. There are two classes of system to be considered, according as the strains are due principally to an externally applied load or to the mass of the structure itself.

74

Load Controlling

Here the structure weight is assumed to be negligible compared with the applied load, as is usually the case where a constructional material of high strength is used.   The similarity criteria depend upon whether deformation is in the elastic range below the yield point of the material, or in the plastic range above the yield point.   In the elastic range, the criterion of similarity is

$$\gamma = \frac{F}{EA} = \text{const} \tag{7-1}$$

where γ = strain
F = applied force
E = elastic modulus
A = stressed area

Using boldface to denote ratios of corresponding quantities, Eq. (7-1) gives the scale equation

$$\mathbf{F} = \mathbf{E}\mathbf{L}^2 \tag{7-1a}$$

where L = corresponding linear dimension.   In homologous systems,

$$\mathbf{F} = \mathbf{L}^2 \tag{7-1b}$$

In the plastic range, the form of the relation is similar, but the elastic modulus is replaced by the yield stress Y, giving the criterion

$$\gamma = \frac{F}{YA} = \text{const} \tag{7-2}$$

Whence the scale equation

$$\mathbf{F} = \mathbf{Y}\mathbf{L}^2 \tag{7-2a}$$

which for homologous systems reduces again to (7-1b).

Plasticine models have been used to study the severe plastic deformation undergone by metals in forming processes such as punching, forging, wire drawing, and rolling (106).   They are applicable where the metal is non-work-hardening, i.e., where a small elastic deformation is followed by a large plastic deformation without further rise of stress.

Where part of a structure deforms elastically and part deforms plastically but without rupture, static similarity requires that both the elastic and the plastic criteria shall be satisfied, i.e.,

$$\gamma = \frac{F}{EA} = \text{const} \qquad \text{and} \qquad \frac{E}{Y} = \text{const}$$

It is seldom feasible to employ two different materials of construction whose elastic moduli and yield points are in the same ratio, and therefore, where a structure is subject to both elastic and plastic deformation, the model should be constructed of the same material as the prototype (homologous systems).

Mass Controlling

A solid structure suffers a certain deformation due to its own weight, and the amount may be determined by observations on suitable models. In the case of a rotating body, the centrifugal force is a function of the mass, and a model may be used to predict the point of failure. There are four similarity criteria, depending upon whether the mass force is gravitational or centrifugal and whether the resulting deformation is elastic or plastic.

Considering first elastic deformation under gravity, the force due to structure weight is proportional to $\rho L^3 g$, where ρ = density of material, L = linear dimension, and g = acceleration of gravity. Inserting this expression into Eq. (7-1) and putting $A = L^2$,

$$\gamma = \frac{\rho L g}{E} = \text{const} \tag{7-3}$$

Where the only force acting upon the mass is that of gravity, local variations in g may be neglected and $\mathbf{g}$ put $\approx$ unity. This gives the scale equation

$$\mathbf{L} = \frac{\mathbf{E}}{\varrho} \tag{7-3a}$$

There is no scale equation for homologous systems, since geometrically similar structures made of the same material do not suffer geometrically similar distortions. The weight of the structure varies as L^3, whereas the cross section that carries it varies as L^2.

For plastic deformation under gravity the corresponding criterion is

$$\gamma = \frac{\rho L g}{Y} = \text{const} \tag{7-4}$$

giving the scale equation

$$\mathbf{L} = \frac{\mathbf{Y}}{\varrho} \tag{7-4a}$$

Hence, under a mass-controlled static regime subject to gravitational forces only, the model must be constructed of a material having a lower ratio of either elastic modulus or yield stress to density (according as the elastic or plastic range is under investigation). The scale ratio is then fixed by Eq. (7-3a) or (7-4a). (Rubber and plasticine might, for example, be employed as constructional materials for the elastic and plastic ranges, respectively.)

The effect of *centrifugal force* on a rotating solid body is an example of a static regime in a dynamic system. Provided that the speed of rotation is constant, the body can be considered as though it were stationary and subject to constant radial forces. Centrifugal systems in which there is no radial motion are analogous to static gravitational systems except for the direction of the mass forces and the fact that in the centrifugal system they vary with velocity and radius.

In the case of centrifugal force, the virtual acceleration g' varies with the speed of rotation N; consequently, similarity between homologous systems is possible. $g' = N^2L$, whence the total centrifugal force is proportional to ρN^2L^4. Substituting in Eq. (7-1) gives the similarity criteria for elastic deformation:

$$\gamma = \frac{\rho N^2L^2}{E} = \text{const} \tag{7-5}$$

and the scale equation

$$\mathbf{N} = \frac{1}{\mathbf{L}}\sqrt{\frac{\mathbf{E}}{\varrho}} \tag{7-5a}$$

or, for homologous systems,

$$\mathbf{N} = \frac{1}{\mathbf{L}} \tag{7-5b}$$

For plastic deformation and ultimate failure under centrifugal force, the corresponding equations are

$$\gamma = \frac{\rho N^2L^2}{Y} = \text{const} \tag{7-6}$$

$$\mathbf{N} = \frac{1}{\mathbf{L}}\sqrt{\frac{\mathbf{Y}}{\varrho}} \tag{7-6a}$$

For homologous systems, the scale equation reduces to (7-5b).

Hence, for example, if a model centrifuge bowl is found to fail at a certain speed of rotation, the speed at which a geometrically similar prototype constructed of the same material may be expected to fail is given by Eq. (7-5b).

All of the above criteria assume that the material of construction is homogeneous and structureless. In practice, most materials have a granular microstructure, and for strict similarity it would be necessary for the ratio of coresponding grain sizes in the model and prototype to be equal to the scale ratio. Provided that the smallest cross section in the model is large compared with the grain size and surface irregularities of the material, this factor may be neglected. It does, however, set a minimum size to models constructed of coarse-grained materials.

Mixed Regime

A static regime is mixed when deformation is due partly to structure weight or mass and partly to an externally applied load, neither factor being small enough to be neglected. Under these conditions, strict similarity between model and prototype is not possible. An approximation to similarity may be attained in the special case where the applied load can be assumed to be distributed in the same manner as the structure weight, e.g., when a uniformly distributed load is applied to a beam of constant cross section. Then we can write for the total load $F + mg$, where $F =$ applied load, $m =$ mass of structure. For elastic deformation, Eq. (7-1) becomes

$$\gamma = \frac{F + mg}{EA} = \text{const} \qquad (7\text{-}7)$$

Or, expressing both applied load and mass load as weights ($F = W$, $mg = w$),

$$\mathbf{W} + \mathbf{w} = \mathbf{E}\mathbf{L}^2 \qquad (7\text{-}7a)$$

where
$$\mathbf{W} + \mathbf{w} = \frac{W' + w'}{W + w}$$

From this scale equation, knowing $\mathbf{L}$, $\mathbf{E}$, and ϱ, one can calculate the magnitudes of the applied loads for similar total deformations. If W' and w' are the applied and gravitational loads, respectively, on the prototype structure, then

$$W = \frac{1}{\mathbf{L}^2}\left[\frac{1}{\mathbf{E}}\,W' + w'\left(\frac{1}{\mathbf{E}} - \frac{1}{\varrho\mathbf{L}}\right)\right] \qquad (7\text{-}7b)$$

Example. A steel bridge weighing 10,000 lb is to carry a distributed load of 20,000 lb. A model is constructed in aluminum to a scale ratio of 10. What load must be applied to the model in order to establish static similarity?

$$W' = 20,000 \qquad w' = 10,000 \qquad L = 10$$
$$\varrho = \frac{7.7}{2.7} = 2.85 \qquad E = \frac{30 \times 10^6}{10.3 \times 10^6} = 2.91$$

whence $W = 68 + 30.5 = 98.5$ lb.

68 lb is the weight corresponding to a 20,000-lb load on the full scale, and an additional distributed load of 30.5 lb is required to simulate the extra deformation due to structure weight alone in the large bridge.

Dynamic Regime

Dynamic systems may consist either of solid bodies only or of solids and fluids together. There are no purely fluid systems because every fluid is somewhere in contact with a solid boundary and the reaction between the boundary and the fluid influences the behavior of the system. Although for convenience one may refer to fluid systems, the presence of a solid envelope, or boundary, is always implied.

Machines and mechanical movements include many systems which may be treated as being composed of solid members only (the effect of the atmosphere upon their motions being generally negligible.) Numerous different types of machine are employed in chemical and metallurgical processes, for example, pulverizers, centrifuges, presses, elevators, and conveyors; but their structural and kinematic design is considered to be the province of the mechanical engineer rather than of the chemical engineer, and model theory as applied to machine design will not be discussed here.

Fluid Systems

Similarity criteria for fluid systems are derived from the equations of motion, supplemented by dimensional analysis. The fundamental differential equations governing the motion of a viscous fluid are the Navier-Stokes equations, which were discussed in Chap. 5. For geometrically similar envelopes (solid-fluid surfaces), the equations may be written in the generalized form

$$\frac{\Delta p}{L} = \phi' \left(\frac{\rho v^2}{L}, \frac{\mu v}{L^2}, \rho g \right) \tag{7-8}$$

where Δp = pressure drop
L = corresponding linear dimension
v = fluid velocity
ρ, μ = fluid density and viscosity
g = acceleration of gravity

Dividing across by $\rho v^2/L$ and rearranging,

$$\frac{\Delta p}{\rho v^2} = \phi \left(\frac{\rho v L}{\mu}, \frac{v^2}{Lg} \right) \tag{7-9}$$

This is the generalized dimensionless equation describing the motion of a viscous fluid. It is not complete, however, since the Navier-Stokes equations do not take into account the effects of surface or interfacial tension, which become important when one fluid is dispersed as droplets in another. Dimensional analysis indicates that these effects are functions of another dimensionless group, $\rho v^2 L/\sigma$, where σ = interfacial tension. Hence, the complete dimensionless equation for fluid motion is

$$\frac{\Delta p}{\rho v^2} = \phi\left(\frac{\rho v L}{\mu}, \frac{v^2}{Lg}, \frac{\rho v^2 L}{\sigma}\right) \tag{7-10}$$

The group on the left is the pressure coefficient; those on the right are, respectively, the Reynolds number, Froude group, and Weber group.

For homologous systems containing the same fluid or fluids in geometrically similar envelopes of different absolute magnitudes, the three dimensionless groups on the right of Eq. (7-10) are manifestly incompatible. The required relations between corresponding velocities and corresponding lengths are:

Reynolds number: $v \propto \dfrac{1}{L}$

Froude group: $v \propto \sqrt{L}$

Weber group: $v \propto \dfrac{1}{\sqrt{L}}$

By the use of fluids with different physical properties in the two systems and the choice of appropriate linear and velocity scale ratios, it is theoretically possible to satisfy any two of the three dimensionless criteria simultaneously. This is sometimes done in model experiments intended to throw light on the dynamics of a particular piece of equipment, but it is not possible in the case of pilot plants where the same materials must be processed on the small and the large scale. If the results of pilot-plant experiments are to be scaled up with reasonable confidence, it is desirable that the reactions and physical operations shall be conducted under conditions such that each reaction rate is predominantly controlled by one particular dimensionless group. This leads to three principal subdivisions of the fluid dynamic regime according as the Reynolds number, Froude group, or Weber group predominates. These three kinds of fluid dynamical regime may be briefly described as viscosity-controlled, gravity-controlled, and surface-tension-controlled (more correct terms might be inertia-viscosity-controlled, inertia-gravity-controlled, and inertia-surface-tension-controlled).

Viscosity Control

The viscosity-controlled dynamical regime is one of the most important in chemical engineering. Many unit operations are carried out in closed systems under conditions of forced convection, and in these circumstances, not only fluid friction, but also heat- and mass-transfer processes conform to a dynamical regime controlled by the kinematic viscosity of the fluid. Rate equations for the regime have been determined empirically for various types and shapes of solid envelope: but, in systems of complicated or unusual form, it is still necessary to resort to model experiments in order to predict performance on the large scale.

For this regime, the generalized dimensionless equation of motion reduces to

$$\frac{\Delta p}{\rho v^2} = \phi\left(\frac{\rho v L}{\mu}\right) \tag{7-11}$$

Using, as before, boldface type to denote ratios of corresponding quantities, Eq. (7-11) gives the following set of scale equations:

$$\mathbf{v} = \frac{\mathbf{\mu}}{\mathbf{\rho L}} = \frac{\mathbf{\nu}}{\mathbf{L}} \tag{7-11a}$$

$$\mathbf{q} = \frac{\mathbf{\mu L}}{\mathbf{\rho}} = \mathbf{\nu L} \tag{7-11b}$$

$$\mathbf{\Delta p} = \frac{\mathbf{\mu}^2}{\mathbf{\rho L}^2} \tag{7-11c}$$

$$\mathbf{P} = \frac{\mathbf{\mu}^3}{\mathbf{\rho}^2\mathbf{L}} \tag{7-11d}$$

where q = total volumetric rate of flow (e.g., in cfs or gpm)
   ν = kinematic viscosity
   P = power consumption

For homologous systems, $\mathbf{\rho}$ and $\mathbf{\mu}$ become equal to unity, and the scale equations reduce to

$$\mathbf{v} = \mathbf{P} = \frac{1}{\mathbf{L}} \tag{7-11e}$$

$$\mathbf{q} = \mathbf{L} \tag{7-11f}$$

$$\mathbf{\Delta p} = \frac{1}{\mathbf{L}^2} \tag{7-11g}$$

Note. In the above equations, $\mathbf{v}$ is the ratio of fluid velocities at corresponding points in the two systems or, more usually, the ratio of mean velocities over corresponding cross sections of the fluid path. In the case of vessels stirred by a revolving paddle or impeller, performance is correlated in terms of the peripheral speed of the paddle = $\pi n d$, where n = rps, d = paddle diameter. In such systems, the linear dimension is usually equated to d, and the modified Reynolds number of stirred vessels become $\rho n d^2/\mu$. Dynamical similarity in such systems is treated more fully under Mixing Equipment (Chap. 14).

Equations (7-11a) and (7-11e) define corresponding velocities in geometrically similar systems under a viscosity-controlled dynamical regime, and Eqs. (7-11b) and (7-11f) give the corresponding volumetric flows. Under these conditions, given geometrically similar envelopes, the fluid-flow patterns will also be geometrically similar. The pressure drop and power consumption for a large-scale system can then be predicted from model experiments using Eqs. (7-11c) [(7-11g)] and (7-11d) [(7-11e)], respectively.

Heat and mass transfer by forced convection follow equations that are in every respect analogous; so they may conveniently be treated together. The generalized equations of motion are:

For heat transfer:
$$\frac{hL}{k} = \phi\left(\frac{\rho vL}{\mu}, \frac{c_p \mu}{k}\right) \qquad (7\text{-}12)$$

For mass transfer:
$$\frac{KL}{D} = \phi\left(\frac{\rho vL}{\mu}, \frac{\mu}{\rho D}\right) \qquad (7\text{-}13)$$

where h = heat-transfer coefficient
   k = thermal conductivity of fluid
   c_p = specific heat of fluid at constant pressure
   K = mass-transfer coefficient
   D = diffusion coefficient

The groups hL/k, $c_p\mu/k$, and $\mu/\rho D$ are the Nusselt, Prandtl, and Schmidt or Colburn groups, respectively. KL/D has been called the Sherwood group.

In general, we are more interested in the total quantity of heat or matter transferred in a given apparatus in unit time than in the value of the transfer coefficients. The extensive equations corresponding to (7-12) and (7-13) are

$$\frac{H}{kL\,\Delta T} = \phi\left(\frac{\rho vL}{\mu}, \frac{c_p\mu}{k}\right) \qquad (7\text{-}14)$$

$$\frac{m}{DL\,\Delta C} = \phi\left(\frac{\rho vL}{\mu}, \frac{\mu}{\rho D}\right) \qquad (7\text{-}15)$$

where H and m = total amounts of heat and matter, respectively, transferred in unit time
   ΔT = temperature difference
   ΔC = concentration difference (mass basis)

Both Prandtl and Schmidt groups contain only the physical properties of the fluid; hence, for homologous systems they are constant. For systems that are not homologous, the Prandtl or Schmidt group,

respectively, will in general be different, but the difference is often
small enough to be neglected.   For example, the Prandtl number for
most gases does not differ greatly from 0.75.   Under these conditions
(and in geometrically similar envelopes), the criterion for similarity
reduces to constant Reynolds number.   The velocity, pressure, and
power relations of Eqs. (7-11a) to (7-11g) apply, and Eqs. (7-14) and
(7-15) yield the following additional scale equations:

$$h = \frac{k}{L} \tag{7-15a}$$

$$K = \frac{D}{L} \tag{7-15b}$$

$$H = kL \, \Delta T \tag{7-15c}$$
$$m = DL \, \Delta C \tag{7-15d}$$

Or, in the case of homologous systems,

$$h = K = \frac{1}{L} \tag{7-15e}$$

$$H = m = L \tag{7-15f}$$

True dynamical similarity in homologous viscosity-controlled systems requires that the fluid velocity, pressure drop, and power consumption shall all be higher in the model than in the prototype.   This is
often impracticable, and where there is a large scale reduction, it may
bring the model system into the region of sonic velocities, where different similarity criteria obtain.   One may therefore have to be satisfied
with a condition of extrapolated similarity in which heat- and mass-
transfer coefficients are caused to be of approximately the same magnitude in both model and prototype (see Chap. 8).

Gravity Control

In gaseous systems, the effect of gravity is negligible except where
there are large density differences due to either temperature or concentration gradients.   In liquid systems, gravity control is found where a
free liquid surface is subject to disturbance, as with liquid sprays or
jets in air or waves and vortices at a liquid surface.   For such phenomena, the Froude group is the criterion of similarity, and the dimensionless equation reduces to

$$\frac{\Delta p}{\rho v^2} = \phi \left(\frac{v^2}{Lg} \right) \tag{7-16}$$

Using boldface type to denote ratios of corresponding quantities and putting $\mathbf{g} = 1$, Eq. (7-16) yields the following scale equations:

$$\mathbf{v} = \sqrt{\mathbf{L}} \tag{7-16a}$$
$$\mathbf{q}_v = \mathbf{L}^{2.5} \tag{7-16b}$$
$$\mathbf{\Delta p} = \mathbf{\varrho L} \tag{7-16c}$$
$$\mathbf{P} = \mathbf{\varrho L}^{3.5} \tag{7-16d}$$

where q_v = total volumetric flow (e.g., in cfs) and P = power consumption. For homologous systems, $\varrho = 1$.

In geometrically similar vessels or flow passages, if any one of the scale equations is satisfied, then the others will also hold. The power consumption P refers only to the power spent in raising material against the force of gravity, and not that absorbed in fluid friction.

If the net gravitational force is proportional to the difference between two densities, as when there are two immiscible liquids, the density difference $\Delta\rho$ is substituted for ρ in the dimensionless and scale equations.

The conditions for similarity under a gravity-controlled dynamic regime allow of reasonable velocities and power consumptions on the small scale. Hence, it is seldom necessary to resort to extrapolated similarity.

The classic application of similarity principles to a gravity-controlled dynamic regime was Froude's use of model hulls to predict the resistance due to wave formation by steamships. In chemical engineering, the Froude group enters where there is a free liquid surface in which ripples or vortices are formed and it is desired to examine their shape and behavior. Thring has studied the disturbance of the molten metal in a side-blown Bessemer converter using models in which mercury or water was subjected to a cold-air blast (64). On purely dimensional grounds, it would follow that for similarity both the Froude group and the ratio of air to liquid density ρ_a/ρ_l should be constant. Thring found that he could combine these into a "modified Froude number,"

$$\frac{\rho_a}{\rho_l}\frac{v_a^2}{Lg}$$

Provided this group was kept constant, the surface disturbance of the liquid conformed to a similar pattern even with widely different values of ρ_a/ρ_l.

Surface-tension Control

When two immiscible liquids are agitated together, one becomes dispersed as droplets in the other. Provided the viscosities of the liquids are low and their densities are nearly equal, the effect of viscosity and gravity upon the flow pattern may be neglected.   The degree of dispersion and range of droplet sizes then depend only upon the geometry of the system, vigor of agitation, and interfacial tension of the two liquids.   The general mechanism of dispersion is that through turbulence globules of the disperse phase are set spinning, and these by centrifugal force are caused to break up into smaller globules until an order of magnitude is reached at which centrifugal forces are balanced by surface-tension forces holding the globules together.

If a dispersed globule of diameter L is spinning with a maximum peripheral velocity v, then the total centrifugal force acting on the globule is proportional to $\rho L^2 v^2$ and the total surface-tension force is proportional to σL, where ρ, σ are, respectively, the density of the disperse phase and the interfacial tension.   For dynamic similarity, the criterion is a constant ratio of centrifugal to interfacial forces, whence

$$\frac{\rho v^2 L}{\sigma} = \text{const} \qquad (7\text{-}17)$$

This dimensionless group is known as the *Weber group*, and its constancy defines the condition under which two-phase liquid systems in geometrically similar envelopes will form geometrically similar dispersions.   For homologous systems, the corresponding scale equations are

$$\mathbf{v} = \frac{1}{\sqrt{L}} \qquad (7\text{-}17a)$$

$$\mathbf{N} = \frac{1}{L^{3/2}} \qquad (7\text{-}17b)$$

$$\mathbf{s} = \frac{1}{L} \qquad (7\text{-}17c)$$

where N = angular velocity, or rpm of stirrer and s = specific interfacial surface (per unit volume).   The ratio of power consumptions cannot be calculated from the Weber group.   It depends upon the Reynolds number with respect to the continuous phase, and in homologous systems equal Weber and Reynolds numbers are impossible.   Equation (7-17c) holds for any geometrically similar systems, whether homologous or not.

The scale equations derived from the Weber criterion are, in practice, of limited utility.   Usually, the purpose of dispersing one liquid in another is to promote a physical or chemical reaction between them.   When the reaction rate is controlled by the rate of mass transfer between the phases, there exist incompatible regimes under which true similarity is impossible. The reaction rate will depend partly upon the specific surface, which is a function of the Weber group, and partly on the over-all mass-transfer coefficient, which is a function of the Reynolds number. In such systems, the effect of increased velocity is twofold: it increases both the total interfacial area and the mass-transfer coefficient per unit area. Hence, the apparent Reynolds index is exceptionally high. For mass transfer between immiscible liquids in stirred vessels, Hixson and Smith obtained values varying from 3.8 to 5.0 depending upon the geometry of the system. (The *Reynolds index* is the exponent x in the relation: Reaction rate $\propto v^x$; see Chap. 6. In mixing theory, it is generally called the *mixing slope*.)

Where the Reynolds index for a given system and geometrical form of envelope has been determined by experiment, the conditions for equal reaction rates, or equal mass-transfer rates per unit volume, can be calculated as in the viscosity-controlled regime from Eqs. (7-16) and (7-17) and the corresponding scale equations. This allows of a condition of pseudo similarity which is probably the best that can be aimed at under mixed viscosity and surface-tension control.

The Weber group is not incompatible with the criterion for chemical similarity, and where there is a slow chemical reaction between immiscible liquids, so that the effect of mass-transfer resistance may be neglected, a simultaneous equating of the chemical and Weber groups allows similarity to be established. This case is considered later in connection with *chemical similarity*.

It would appear that Eqs. (7-17) to (7-17c) should apply to liquid-liquid dispersers and colloid mills, allowing the performance of geometrically similar mills to be correlated. The authors are not aware of any published information on this application.

Thermal Regime

A thermal regime exists where the criteria of similarity cannot be formulated solely in terms of mass, length, and time, even in homologous systems, but where temperatures and temperature differences have also to be introduced into the dimensionless groups.

There are five processes which may take place in a heat-transfer system in which there is no chemical action: the transport of heat by bulk movement of material, and heat transfer by conduction, forced convection, natural convection, and radiation. The dimensionless equation for natural convection was obtained by dimensional analysis in Chap. 4 [Eq. (4-4)]. Corresponding equations for the other processes were derived from the rate equations in Chap. 5 [Eqs. (5-24), (5-29), (5-37), (5-40)]. For single-phase fluid systems in which the only effect of gravity is to cause buoyancy, all these equations may be combined into a single generalized dimensionless equation for the thermal regime as a whole,

$$\frac{H}{kL\,\Delta T} = \frac{hL}{k} = \phi\left(\frac{\rho v L}{\mu}, \frac{c_p \mu}{k}, \frac{\beta g\,\Delta T\,L^3\rho^2}{\mu^2}, \frac{\rho c_p v}{\sigma e T^3}, \frac{T_a}{T_r}\right) \qquad (7\text{-}18)$$

where H = net rate of heat transfer for whole system
            $= dQ/dt$
        h = heat-transfer coefficient
        L = linear dimension
k, ρ, μ, c_p, β = thermal conductivity, density, viscosity, specific heat at constant pressure, volumetric coefficient of thermal expansion of fluid
        v = fluid velocity
      ΔT = temperature difference
    T_r, T_a = absolute temperatures of corresponding points on radiating and absorbing surfaces
        e = combined emissivities of radiating and absorbing surfaces
        σ = Stefan-Boltzmann constant

The term on the left is the Nusselt group; the first three between parentheses are, respectively, the Reynolds, Prandtl, and Grashof groups; the fourth is a group used by Thring as a measure of the ratio between bulk transport of heat and radiation (64), and which will be called the *radiation group*. The Peclet group $\rho c_p v L/k$, obtained by multiplying together the Reynolds and Prandtl groups, may be substituted for the latter in Eq. (7-18).

The Reynolds, Grashof, and radiation groups are in general incompatible, although by a special choice of temperatures and materials it may sometimes be possible to keep two of them the same in model and prototype systems. The Reynolds and Prandtl groups contain no term measuring the temperature of the system. Where these criteria alone control the heat-transfer rate, the regime is dynamic and not thermal, and the conditions for similarity have already been given [Eqs. (7-11) to (7-15f)]. This leaves only two strictly thermal regimes, controlled, respectively, by natural convection and radiation.

Natural-convection Control

The generalized dimensionless equation here reduces to

$$\frac{H}{kL\,\Delta T} = \phi\left(\frac{\beta g\,\Delta T\,L^3\rho^2}{\mu^2}, \frac{c_p \mu}{k}\right) \qquad (7\text{-}19)$$

The first requirement for thermal similarity, as in the case of forced convection, is that the Prandtl groups in the two systems shall be approximately equal. On this assumption, and putting $\mathbf{g}$ = unity, the following scale equations are obtained:

$$\mathbf{\Delta T} = \frac{\mathbf{u}^2}{\beta \varrho^2 \mathbf{L}^3} \tag{7-19a}$$

$$\mathbf{H} = \frac{\mathbf{k} \mathbf{u}^2}{\beta \varrho^2 \mathbf{L}^2} \tag{7-19b}$$

$$\mathbf{h} = \frac{\mathbf{k}}{\mathbf{L}} \tag{7-19c}$$

For homologous systems:

$$\mathbf{\Delta T} = \frac{1}{\mathbf{L}^3} \tag{7-19d}$$

Equations (7-19a) and (7-19d) lay down the value of ΔT necessary for thermal similarity, and the other scale equations give the resulting ratios of total heat flow and heat-transfer coefficients. It is evident that in homologous systems where the scale ratio is large the temperature difference would have to be impracticably high in the model in order to simulate a reasonable value in the prototype. By choosing a fluid of higher viscosity for the model system and selecting the scale ratios and temperature difference accordingly, ΔT may be brought within a practicable range. Hence, in a pilot plant, where homologous systems are essential, rates of heating under a thermal regime controlled by natural convection cannot as a rule be strictly simulated, and the best that can be achieved is a condition of partial or extrapolated similarity (see Chap. 8). In a model experiment designed to predict heating rates with a particular geometrical arrangement of surfaces, the use of a more viscous fluid in the model may permit true thermal similarity to be attained.

Radiation Control

In a continuous-flow system in which the only appreciable heat-transfer processes are conduction, radiation, and bulk transport of heated material, the generalized dimensionless equation reduces to that derived in Chap. 5 [Eq. (5-45)],

$$\frac{H}{kL\,\Delta T} = \frac{hL}{k} = \phi \left(\frac{\rho c_p v}{\sigma e T^3}, \frac{T_1}{T_2}, \frac{\rho c_p L v}{k} \right) \tag{7-20}$$

For geometrically similar and fully homologous systems of different sizes, the radiation and Peclet groups are incompatible, since in the former v is constant, while in the latter it is inversely proportional to L. Hence, even when convection has been eliminated, there is still a mixed regime.

In many of the radiating systems met with in practice, such as furnaces or high-temperature reactors, convective effects are negligible owing to the high kinematic viscosity of hot gases, and conduction is important only in so far as it affects the loss of heat through the boundary walls. In such cases, strict similarity with respect to conduction is not necessary provided that the wall thickness of the model can be fixed independently of the scale ratio in other respects. Similarity with respect to radiation and bulk transport may then be attained.

Consider two reaction vessels which are geometrically similar in their internal dimensions but not necessarily in wall thickness. Let ΔT represent the temperature difference across the walls and δT the temperature change undergone by the fluid in passing through the vessel. Call the wall thickness w. The rate of bulk transport of heat is then given by $\rho c_p v L^2\ \delta T$, while the rate of conduction through the walls is proportional to $kL^2\ \Delta T/w$. Hence, the ratio of heat transported to heat lost by conduction is

$$\frac{\rho c_p v\ \delta T\ w}{k\ \Delta T}$$

This group is in effect a modified Peclet group, on the assumption that thermal conduction in the fluid phase is negligible. The wall thickness w having been specified independently of L, the Nusselt group must be written $Hw/kL^2\ \Delta T$. Dividing by the modified Peclet group to eliminate conductivity from the left-hand side of the equation,

$$\frac{H}{\rho c_p v L^2\ \delta T} = \phi\left(\frac{\rho c_p v}{\sigma e T^3}, \frac{T_1}{T_2}, \frac{\rho c_p v\ \delta T\ w}{k\ \Delta T}\right) \tag{7-21}$$

The radiation and modified Peclet groups are no longer incompatible. For homologous fluid-phase systems in which corresponding temperatures are equal, T_1/T_2 is constant, and $\delta T = 1$. The scale equations for similarity are then

$$\mathbf{v} = 1 \tag{7-21a}$$

$$\frac{\mathbf{k}\,\mathbf{\Delta T}}{\mathbf{w}} = 1 \tag{7-21b}$$

$$\mathbf{H} = \mathbf{L}^2 \tag{7-21c}$$

In words, both the fluid velocities and the conducted heat loss per unit area of wall must be made equal in model and prototype. The total quantity of heat transferred in unit time will then be proportional to the square of the linear dimension, i.e., to the volumetric rate of flow. If the walls are made of the same material and subject to the same external temperature, then

$$\mathbf{w} = 1 \qquad (7\text{-}21d)$$

Both wall thicknesses should be the same.

The foregoing discussion anticipates somewhat the treatment of wall effects in Chap. 9. It was necessary here because in radiating systems the wall effects are often inseparable from the main process.

Chemical Regime

A chemical regime prevails when the over-all rate of change is controlled by a chemical-reaction velocity, which consequently appears in one or more of the dimensionless criteria. The application of similarity principles to chemical-reaction systems has been discussed by Damköhler (7, 8), Edgeworth Johnstone (18), Hurt (17), Hulbert (16), Dodd and Watson (12), Bosworth (1 to 4) and Thring (64, 258).

The scaling up or down of a chemical regime becomes a problem only when the main reaction is accompanied by various side reactions which reduce the yield and which may behave differently from the main reaction when the size of the system is changed. It is easy to find two chemically reacting systems in which the main reactions are analogous (e.g., of the same order and having similar temperature coefficients), but it is very difficult to find systems in which all the side reactions are also known to be analogous. Therefore, in effect, the application of similarity principles to the chemical regime is limited to homologous systems.

For a continuous-flow system Damköhler (7) proposed five dimensionless criteria,

$$\underset{\text{I}}{\frac{UL}{av}} = \phi \left(\underset{\text{II}}{\frac{UL^2}{aD}}, \underset{\text{III}}{\frac{qUL}{c_p\rho Tv}}, \underset{\text{IV}}{\frac{qUL^2}{kT}}, \underset{\text{V}}{\frac{\rho vL}{\mu}} \right) \qquad (7\text{-}22)$$

where U = reaction rate expressed as moles of reactant A which react per unit volume and time

a = concentration of reactant A per unit volume

D = diffusion coefficient of A

q = heat generated per mole of A reacting (distinguish from q_v = volumetric rate of flow)

c_p, ρ, k, μ = specific heat at constant pressure, density, thermal conductivity, and viscosity of reaction mixture, respectively

T = temperature of reaction mixture

v = linear velocity of reaction mixture

L = linear dimension

The significance of Damköhler's similarity groups as ratios of process rates are as follows:

I $\dfrac{\text{Chemical reaction}}{\text{Bulk flow}}$

II $\dfrac{\text{Chemical reaction}}{\text{Molecular diffusion}}$

III $\dfrac{\text{Heat liberated}}{\text{Heat transported by bulk flow}}$

IV $\dfrac{\text{Heat liberated}}{\text{Heat transported by conduction}}$

V $\dfrac{\text{Momentum transferred by bulk flow}}{\text{Momentum transferred by viscosity}}$ (Reynolds number)

Damköhler's dimensionless terms may be transformed into more customary groups, eliminating U from the right-hand side of the equation,

$$\frac{UL}{av} = \phi\left(\frac{\mu}{\rho D}, \frac{qa}{c_p \rho T}, \frac{c_p \mu}{k}, \frac{\rho v L}{\mu}\right) \tag{7-23}$$

Thus, the chemical-reaction velocity group on the left is a function of the Schmidt, Prandtl, and Reynolds numbers and a group containing an entropy term.

Equation (7-23) neglects the effect of heat transfer by radiation. Many chemical reactions occur at temperatures such that radiation is an important if not the predominant mode of heat transfer, so that a complete generalized equation should include Thring's radiation group and the ratio of absolute temperatures as introduced into Eq. (7-18). The full equation then becomes

$$\frac{UL}{av} = \phi\left(\frac{\mu}{\rho D}, \frac{qa}{c_p \rho T}, \frac{c\mu}{k}, \frac{\rho c_p v}{\sigma e T^3}, \frac{T}{T_r}, \frac{\rho v L}{\mu}\right) \tag{7-24}$$

This is the generalized rate equation for a continuous-flow chemically reacting system. Certain minor effects such as heat transfer by natural convection have already been neglected, but Eq. (7-24) still contains seven dimensionless groups which cannot all be kept constant when the size of the system is changed. Certain of these groups can in practice be neglected without seriously affecting similarity, but this may introduce an appreciable scale effect, a possibility which has to be borne in mind when a chemical reaction is scaled up or down.

In arriving at a simpler equation, we begin by neglecting the Schmidt group. This is equivalent to assuming that mass transfer by molecular diffusion is negligible compared with transport by eddy diffusion and bulk flow. Such an assumption may not be justified in gaseous systems where the fluid motion is likely to be streamline, but for gases the Schmidt group is approximately constant in any case.

It is often permissible also to disregard the Reynolds number, i.e., to assume that the flow pattern does not significantly influence the chemical reaction. Strictly speaking, this assumption is justified only in two extreme regions, at high Reynolds numbers, where the velocity profile is nearly flat, and in the streamline region, where it becomes parabolic. In both cases, the ratio of mean to maximum velocity is approximately constant, the numerical values being 0.5 in the streamline region and about 0.8 at Reynolds numbers above 10,000. At intermediate Reynolds numbers, this ratio may vary appreciably as between model and prototype, especially if the flow happens to be streamline in the one and turbulent in the other. As a result, the statistical distribution of residence times in the reaction zone will be different for the molecules of the two systems, and this may affect the yield.

The final simplifying assumption is that transverse temperature gradients inside the reacting system are negligible compared with the gradient through the walls of the reaction vessel. This is equivalent to assuming that the pattern of temperature distribution does not significantly influence the chemical reaction, and it is justified only where temperature gradients within the system are relatively small. In practice, the heat flux per unit area of reaction-vessel wall H_w can be artificially varied by devices such as insulation, jacketing, or electrical heating of the reaction vessel. It is therefore an independent variable, and the Prandtl and radiation groups together with the ratio of absolute temperatures may be replaced by a single group qUL/H_w, representing the ratio of heat generated to heat lost or gained through the vessel walls. It is again convenient to eliminate U by dividing by UL/av, whereby the group becomes qav/H_w.

With the above three simplifying assumptions, the generalized dimensionless equation for chemical similarity becomes

$$\frac{UL}{av} = \phi\left(\frac{qa}{c_p\rho T}, \frac{qav}{H_w}\right) \tag{7-25}$$

where H_w = heat flow through reaction-vessel walls per unit area and time.

There are two principal subdivisions of the chemical regime according as the main reaction is homogeneous or heterogeneous. In the first case, the major rate-determining factor is mass action; in the second, it is the extent of surface or interface between the phases.

Mass-action Control (Homogeneous Reactions)

The general equation for the rate of a homogeneous chemical reaction is

$$U = -\frac{da}{dt} = K_n(a_1 a_2 \cdots a_n)F \qquad (7\text{-}26)$$

where $\quad K_n$ = a velocity constant

$a_1 a_2 \cdots a_n$ = molar concentrations of reactants $A_1 A_2 \cdots A_n$, respectively

$\quad n$ = order of the reaction

$\quad F$ = a dimensionless kinetic factor derived from the activity coefficients of the reactants and products

The dimensions of K_n vary with the order of the reaction so that the product $K_n(a_1 a_2 \cdots a_n)$ always has the dimensions $m/L^3 t$ (see Chap. 4 and Appendix 2).

Putting $a = a_1$, Eq. (7-25) becomes for homogeneous reactions

$$\frac{K_n F(a_2 \cdots a_n)L}{v} = \phi\left(\frac{qa_1}{c_p\rho T}, \frac{qa_1 v}{H_w}\right) \qquad (7\text{-}27)$$

For homologous systems, the scale equations are

$$\mathbf{v} = \mathbf{L} \qquad (7\text{-}27a)$$
$$\mathbf{H} = \mathbf{L} \qquad (7\text{-}27b)$$
$$\mathbf{q}_v = \mathbf{L}^3 \qquad (7\text{-}27c)$$
$$\mathbf{J} = \mathbf{L}^3 = \mathbf{V} \qquad (7\text{-}27d)$$

where q_v = total volumetric flow (e.g., in ft³/hr)

$\quad J$ = total heat loss per unit time through the reaction-vessel walls

$\quad V$ = volume of reaction space

The similarity criteria therefore consist in equal residence times in the reaction zone with surface-heat losses per unit area reduced in proportion to the linear scale. This often necessitates a so-called "adiabatic" jacket for the model. Since the flow pattern is assumed not to influence the reaction, geometrical similarity of reaction vessels is not necessary. If the vessels are not geometrically similar, the total heat lost through the walls must be reduced in proportion to the volume of the reaction space [Eq. (7-27d)].

In continuous chemical reactions, the rate of flow of reactants is often stated in terms of space velocity, i.e., volumes of reactants per volume of reactor space per unit time. The space velocity is the reciprocal of the residence time in the reaction zone and is proportional to v/L or q_v/L^3. Equations (7-27a) and (7-27c) are therefore equivalent to specifying equal space velocities on the small and large scale.

Surface Control (Heterogeneous Reactions)

The over-all rates of many heterogeneous chemical reactions are controlled by the rates at which reacting substances are brought together or products removed at the interface. Such reactions are subject to a dynamic regime and not a chemical one.

Apart from dynamic factors, the velocity of a heterogeneous chemical reaction is influenced by the interfacial area between the phases, the extent to which reactants and products are absorbed at the interface, and any specific catalytic activity that the interface may possess. The reaction velocity is represented by a generalized equation of the form

$$U = -\frac{da}{dt} = \phi[K_n F(a_1 a_2 \cdots a_n)\alpha s] \tag{7-28}$$

where K_n = velocity constant for a reaction of the nth order
$\quad\quad F$ = dimensionless kinetic factor as for homogeneous reactions
$\quad\quad s$ = specific surface or interface per unit volume
$\quad\quad \alpha$ = a dimensionless factor proportional to the catalytic activity of the surface or interface

The dimensions of K_n vary for reactions of different orders so that the product $K_n(a_1 a_2 \cdots a_n)s$ always has the dimensions $m/L^3 t$.

For geometrically similar heterogeneous systems, s varies as $1/L$, and Eq. (7-25) becomes

$$\frac{K_n F(a_2 \cdots a_n)\alpha}{v} = \phi\left(\frac{qa}{c_p \rho T}, \frac{Qav}{H_w}\right) \tag{7-29}$$

For homologous heterogenous systems in which surface activities are equal (α constant), the scale equations are

$$\mathbf{v} = \mathbf{H} = 1 \tag{7-29a}$$
$$\mathbf{q}_v = \mathbf{L}^2 \tag{7-29b}$$

For example, in geometrically similar catalytic reactors in which similarity extends to the dimensions of the catalyst grains, the corresponding fluid velocities in model and prototype are equal, and heat losses per unit area of external surface are also equal. No special jacketing of the model reactor is required, and any thermal insulation should be of the same thickness as in the prototype.

Bosworth (3) has pointed out that by varying the activity of the catalyst it is theoretically possible to satisfy simultaneously both the chemical-similarity criteria and the Reynolds number. If α is independently variable, Eq. (7-29a) becomes

$$\mathbf{v} = \mathbf{H} = \alpha \tag{7-29c}$$

For equal Reynolds numbers,

$$\mathbf{v} = \frac{1}{\mathbf{L}} \tag{7-11e}$$

whence, to satisfy both equations,

$$\alpha = \frac{1}{\mathbf{L}} \tag{7-29d}$$

This relation would appear to be of limited utility since economic factors require that the large-scale catalyst shall have a high activity, and it is unlikely that the model catalyst could be $\mathbf{L}$ times as active.

In terms of space velocity, the scale equation for geometrically similar grain or pore sizes becomes

$$\mathbf{S} = \frac{1}{\mathbf{L}} \tag{7-29e}$$

where $\mathbf{S}$ is the ratio of space velocities. The space velocity on the small scale is higher than on the large scale.

In practice, it is often inconvenient to scale down the grain or pore size of a heterogeneous system. It may be difficult to vary the grain size of a given solid catalyst without altering its activity, in which case it is better for the catalyst used in the pilot plant to be identical with that to be used on the large scale. Where the grain or pore size remains constant, there is a particular advantage in having the small-scale reactor as an element of the prototype, i.e., with the same length of travel of the reactants. Mean linear velocities and Reynolds numbers (based on pore diameter) are then the same in both systems, and provided that surface heat losses are suitably controlled, a closer approach to similarity is possible than with a model.

For the general case (covering both elements and models) in which over-all linear dimensions and specific surface are independent of one another, the generalized equation becomes

$$\frac{K_n F(a_2 \cdots a_n)\alpha s L}{v} = \phi\left(\frac{q a_1}{c_p \rho T}, \frac{q a_1 v}{H_w}\right) \tag{7-30}$$

Where s is constant, the scale equations are identical with those for a homogeneous reaction [Eqs. (7-27a) to (7-27d)] and the requirements for similarity are equal space velocities and equal surface-heat losses per unit volume and time.

Mixed Regime

There are three principal ways of dealing with a mixed regime:

1. Let the small-scale apparatus be an element of the prototype rather than a scale model (see Chap. 3), wall effects being neutralized as far as possible by the methods described in Chap. 9.

2. Change the operating conditions so that one regime becomes predominant (Chap. 6). It is important to make sure that the same regime will control the reaction rate on both the small and the large scale.

3. Calculate the effect of one rate-controlling factor, and experiment under conditions of similarity with respect to the other (e.g., Froude's determinations of wave drag using model ship's hulls).

The subject was covered more fully in Chap. 6.

SYMBOLS IN CHAPTER 7

A = reactant
A = stressed area
a = molar concentration of reactant A
C = concentration
c_p = specific heat at constant pressure
D = diffusion coefficient
d = diameter
E = elastic modulus
e = emissivity
F = dimensionless kinetic factor
F = force
g = acceleration of gravity
g' = virtual acceleration due to centrifugal force
H = heat transferred in unit time = dQ/dt
H_w = heat loss per unit area of wall in unit time
h = heat-transfer coefficient
J = total heat loss through reaction-vessel walls in unit time
K = chemical-reaction velocity constant
K = mass-transfer coefficient
k = thermal conductivity
L = linear dimension
m = mass
m = mass transferred in unit time
N = rotational speed, revolutions in unit time
n = number of reactants
P = power consumption
p = pressure

Q = quantity of heat
q = heat of reaction per mole
q = volumetric rate of flow
S = space velocity
s = specific interfacial area (per unit volume)
T = temperature
t = time
U = chemical-reaction velocity, moles per unit volume and time
V = volume of reaction space
v = velocity
W = applied weight
w = structure weight
w = wall thickness
x = Reynolds index
Y = yield point
α = dimensionless catalyst activity factor
β = coefficient of cubical expansion
γ = strain
Δ, δ = difference
μ = viscosity
ν = kinematic viscosity
ρ = density
σ = interfacial tension
σ = Stefan-Boltzmann constant
ϕ = function

CHAPTER 8

EXTRAPOLATION

One of the reasons why model theory has not been more widely applied to pilot plants is that a pilot plant must necessarily process the same materials on the small scale as the production plant will eventually process on the large scale, and under these conditions strict similarity often requires that either the prototype or the model apparatus shall operate under conditions that are impracticable or change the fluid regime. For example, in fully baffled paddle mixers, the criterion of dynamic similarity is equality of the modified Reynolds number. For large-scale ratios, this entails either an uneconomically low stirrer speed in the prototype or a speed in the model so high that cavitation is likely to occur and interfere with the fluid regime. The power input to the model is required to be L times that to the prototype, and the frictional heat evolved per unit volume of fluid is L^4 times that in the prototype. Such high-power inputs are not easily provided in a model, and the heat evolution may appreciably alter the viscosity of the liquid being mixed. Sometimes Reynolds similarity may require fluid velocities on the small scale which approach sonic velocities. The Mach number then becomes important in the model though not in the prototype, the regimes are different, and dynamic similarity is impossible.

Where the prototype apparatus has a multiple structure, it may be feasible to establish similarity by means of a full-scale element or a model element with relatively little scale reduction. This could be done, for example, in the case of a multitubular heater or a packed tower. On the other hand, an apparatus such as a mixer or a chemical stirred tank reactor cannot be considered as made up of multiple elements. Its performance depends upon the flow pattern in the system as a whole. Where dynamic similarity is impracticable, some method is needed whereby model results can be extrapolated to dynamically dissimilar conditions in the prototype.

The subject of extrapolation is best approached by considering the design of a piece of equipment as a process of successive approximation.

Suppose we wish to design a heat exchanger. The first approximation would be to assume a fixed value for the over-all heat-transfer coefficient. Many satisfactory heat exchangers have in fact been designed in this way, but it is a crude approximation and calls for a large factor of safety.

The second approximation is to calculate the individual film coefficients of heat transfer by means of empirical or partly empirical rate equations. This is the basic method for the design of any process plant on paper. The standard rate equations for heat, mass, and momentum transfer are formulated with the aid of dimensional analysis but contain empirical constants. All these rate equations tacitly assume a principle that may be termed the *extended principle of similarity*. As was shown in Chap. 4, the classical principle of similarity is expressed by equations of the form

$$Q = \phi(R, S, \ldots) \tag{8-1}$$

where Q, R, S, . . . are dimensionless groups and ϕ is an unknown function. The extended principle substitutes for ϕ a power function, leading to equations of the form

$$Q = C(R)^x(S)^y \cdots \tag{8-2}$$

where x, y, and C are constants. The physical processes of heat, mass, and momentum transfer have been found to conform fairly closely to power equations of this type.

As a rule, the empirical exponents x and y are only slightly affected by the geometry of the system, while the constant C is very much affected. C is in fact a shape factor. Except in the simplest cases, it cannot be calculated but must be determined by experiment. Hence, rate equations of the type of Eq. (8-2) are applicable only to systems similar in geometrical form to those for which the shape factor was determined. For example, the shape factor for heat transfer in a straight pipe may be quite seriously in error if applied to a coiled pipe.

The third approximation in design is to eliminate the shape factor C by comparing systems of similar geometrical form and taking ratios of dimensionless groups rather than the groups themselves. This entails experimenting with a model, though not necessarily under dynamically or thermally similar conditions. The rate equation for geometrically similar systems then takes the form

$$\frac{Q'}{Q} = \left(\frac{R'}{R}\right)^x \left(\frac{S'}{S}\right)^y \tag{8-3}$$

Not only does the shape factor cancel out, but in homologous systems the physical-property terms in the dimensionless groups cancel out also, leaving a simple relation between ratios of linear dimensions and velocities, of the kind that has been called a scale equation. The only empirical elements remaining are the exponents x and y.

The final and theoretically closest approximation is to eliminate x and y, or the unknown function ϕ, by comparing geometrically similar systems at equal values of the dimensionless criteria. Thus, if

$$R' = R$$
$$\text{and} \qquad S' = S$$
$$\text{then} \qquad Q' = Q \qquad\qquad (8\text{-}4)$$

Equation (8-4) expresses the classical principle of similarity and defines corresponding states for scaling up or down in accordance with that principle. It is these corresponding states which are so often impossible to achieve in practice. By falling back on the extended principle of similarity, and Eq. (8-3), one gains a much greater degree of flexibility, though with some loss of precision due to the slight variability of x and y. This is what is meant by the term *extrapolation*. Though theoretically inferior to strict classical similarity, correct extrapolation by means of Eq. (8-3) is in principle less liable to error than calculation from Eq. (8-2), which contains the highly variable and incalculable shape factor. That is to say that small-scale experiments extrapolated to a geometrically similar but dynamically or thermally dissimilar prototype are likely to be more reliable than values calculated from a rate equation derived from a range of geometrically dissimilar systems.

There are two ways in which the exponents x, y, . . . may be found: either by experiment with the small-scale apparatus or from the literature. Determination by experiment in an apparatus of the actual shape to be used is probably the better method provided a sufficiently wide range of conditions is covered so that it can be verified that the exponent does not vary appreciably. For most of the commoner physical processes and geometrical configurations, however, values of the exponents are available in the literature which are good enough for most purposes. Generally, the experimental exponents are less than unity so that an error in the exponent causes a smaller error in the value of the power function.

The exponent which is most often used in extrapolating similarity conditions is that which has been called the *Reynolds index*, the exponent on the Reynolds number in the empirical rate equations for heat, mass, and momentum transfer by forced convection. This exponent has been determined experimentally by a number of workers for different fluids, operating conditions, and geometrical configurations. Typical values for heat and mass transfer, together with the literature references, are shown in Table 8-1. It will be seen that in geometrically similar systems the Reynolds indices for both transfer processes are substantially equal. In this book, the Reynolds index is denoted by x. Where it is necessary to differentiate between the indices for

TABLE 8-1. TYPICAL REYNOLDS INDICES
Dynamic regime, forced convection, turbulent flow, fixed interface

Transfer process	Flow geometry	Reynolds index	References
Heat.....	Inside pipes and ducts	0.8	161
	Inside annular spaces	0.8	161
	In stirred jacketed vessels	0.67	148
	Across tube banks	0.6	161
	Past spheres	0.6	161
	In stirred vessels with coils	0.5–0.67	148, 170, 207, 208
	Across finned tubes	0.5	161
	Across wires	0.4–0.5	161
Mass.....	In wetted-wall towers	0.8	127
	Over plane surfaces	0.8	141
	Normal to disks	0.65	141
	Through Berl saddles fully wetted	0.65	127
	Across cylinders	0.6	141
	Through Raschig rings fully wetted	0.6	127
	In stirred vessels with solids dissolving	0.6	191
	Through granular solids	0.5–0.6	141
	Past spheres	0.5	141

momentum transfer (friction), heat transfer, and mass transfer, they are distinguished by subscripts, thus: x_f, x_h, x_m. At moderate rates of flow, the Reynolds index for momentum transfer x_f is of the same order of magnitude as the indices for heat and mass transfer, 0.8 to 0.85. At higher velocities, above about four times the critical velocity, x_f diverges from x_h and x_m and approaches unity. The effect of Reynolds number on the flow pattern is then negligible, and the pressure coefficient $\Delta p / \rho v^2$ is almost constant.

Another empirical exponent that is employed for extrapolation is the exponent z in the Lorenz equation for heat transfer by natural convection. The generalized dimensionless equation for heat transfer by natural convection was given in Chap. 4,

$$\frac{hL}{k} = \phi\left(\frac{\beta g\,\Delta T\,L^3\rho^2}{\mu^2}, \frac{c\mu}{k}\right)$$

Lorenz evaluated the function ϕ theoretically and derived an equation of the form

$$\frac{hL}{k} = C\left(\frac{\beta g\,\Delta T\,L^3\rho^2}{\mu^2}\frac{c\mu}{k}\right)^z \tag{8-5}$$

where C is a constant.   The theoretical value of the exponent z is 0.25, but Saunders found that it varies somewhat with the magnitude of the term in parentheses, which is the Grashof number multiplied by

TABLE 8-2. EMPIRICAL INDICES ON (Gr Pr) IN NATURAL CONVECTION
Thermal regime

Range of (Gr Pr)	Fluid motion	Value of z
$>10^9$	Turbulent	0.33
10^8–10^5	Streamline	0.25
$<10^5$	Streamline	~ 0.15

the Prandtl number (Gr Pr).   The empirical values found by Saunders are shown in Table 8-2.   In evaluating the Grashof number, the linear dimension L is taken as the vertical height of a coil or jacket or the mean width of a horizontal heating surface.

In the case of heat transfer by forced convection the rate equation is of the Dittus-Boelter type, and when homologous systems are compared, the physical properties and shape factor cancel out, and Eq. (8-3) reduces to the scale equation

$$h = \frac{v^{x_h}}{L^{1-x_h}} \tag{8-6}$$

For natural convection, the Lorenz equation expressed in ratio form similarly reduces to

$$h = \frac{(\Delta T)^z}{L^{1-3z}} \tag{8-7}$$

The use of these equations in scaling up heat-transfer coefficients to conditions far removed from strict dynamic or thermal similarity is illustrated in Chap. 12.

Mass transfer by forced convection follows an equation similar in form to the Dittus-Boelter and given in Chap. 6 [Eq. (6-3)]. For homologous systems, the scale equation becomes

$$K = \frac{\mathbf{v}^{x_m}}{\mathbf{L}^{1-x_m}} \tag{8-8}$$

This equation applies only where the mass-transfer area is invariable, and this is strictly true only for mass transfer from a fluid to an insoluble solid, e.g., in adsorption. A fluid-fluid interface is always liable to be disturbed and its area changed by the relative movement of the phases. Even in a wetted-wall tower, the surface of the liquid film becomes rippled at moderate gas velocities. Hence, Eq. (8-8) is of less general application for scale-up purposes than Eqs. (8-6) and (8-7).

The equation for fluid friction may be written

$$\frac{\Delta p}{\rho v^2} = C' \left(\frac{\rho v L}{\mu} \right)^{x_f - 1} \tag{8-9}$$

or

$$\frac{\Delta p L}{\mu v} = C' \left(\frac{\rho v L}{\mu} \right)^{x_f} \tag{8-10}$$

leading to the scale equation for homologous systems,

$$\Delta \mathbf{p} = \frac{\mathbf{v}^{1+x_f}}{\mathbf{L}^{1-x_f}} \tag{8-11}$$

This relation also is less generally useful than the heat-transfer relations owing to the variability of x_f between 0.8 and unity. If both model and prototype operate at fluid velocities not less than four times the critical velocity, x_f may be taken as unity and Eq. (8-11) becomes

$$\Delta \mathbf{p} = \mathbf{v}^2 \tag{8-12}$$

Under these conditions, the pressure drop and power consumption can be scaled up for conditions departing from dynamic similarity. The scale equation for power consumption corresponding to (8-12) is

$$P = \mathbf{L}^2 \mathbf{v} \, \Delta \mathbf{p} = \mathbf{v}^3 \mathbf{L}^2 \tag{8-13}$$

In using the extended principle of similarity for scaling up or down it is essential to verify that the regime will remain the same in model and prototype. If, for example, the fluid motion were turbulent in the prototype and streamline in the model, the assumption of a constant Reynolds index would be wildly erroneous.

SYMBOLS IN CHAPTER 8

$$C, C' = \text{shape factors}$$
$$c = \text{specific heat}$$
$$g = \text{acceleration of gravity}$$
$$h = \text{heat-transfer coefficient}$$
$$K = \text{mass-transfer coefficient}$$
$$k = \text{thermal conductivity}$$
$$L = \text{linear dimension}$$
$$p = \text{pressure}$$
$$Q, R, S, \ldots = \text{dimensionless groups}$$
$$T = \text{temperature}$$
$$v = \text{velocity}$$
$$x, y, \ldots = \text{empirical exponents}$$
$$x_f = \text{Reynold index for momentum transfer}$$
$$x_h = \text{Reynold index for heat transfer}$$
$$x_m = \text{Reynold index for mass transfer}$$
$$z = \text{exponent in the Lorenz equation}$$
$$\beta = \text{coefficient of cubical expansion}$$
$$\Delta = \text{difference}$$
$$\mu = \text{viscosity}$$
$$\rho = \text{density}$$
$$\phi = \text{function of} \cdots$$

CHAPTER 9

BOUNDARY EFFECTS

Every experimental system has a boundary which separates it from its surroundings and delimits the variables which are under the experimenter's control. As the system is scaled up, the ratio of boundary surface to internal volume decreases. A beaker 4 in. high by 3 in. in diameter has a surface/volume ratio of 19 ft²/ft³. A geometrically similar pot 4 ft high by 3 ft in diameter has a ratio of only 1.58 ft²/ft³. To raise the surface to volume ratio of the pot to that of the beaker, it would be necessary to depart from geometrical similarity and insert, for example, a coil consisting of about 130 ft of 1-in. tubing. The surface/volume ratio of geometrically similar vessels varies inversely as the linear dimension.

The surfaces or interfaces that influence the performance of process plant are of two kinds: boundary and interior. The boundary surface is normally the wall of the containing vessel. Interior surface may be composed of tubes, Raschig rings, granules, etc., according to the type of plant, and it may also comprise one or more fluid interfaces where there is a polyphase fluid system. A plain vessel has no internal solid surfaces, although it may contain a fluid interface. The area of a fluid interface depends chiefly on the dynamics of the system, while solid-surface areas are a function of geometry alone. The experimental systems of chemical engineering are normally contained in vessels, ducts, or chambers and bounded by solid surfaces. In this chapter, the word *surface* means a solid surface; a liquid surface will be termed an *interface*.

In a small-scale apparatus which constitutes a geometrical element, the interior surface/volume ratio is the same as in the prototype. In a model or model element, the ratio is greater, but the difference is taken account of in the similarity relations. Both elements and models, however, have also a greater boundary surface/volume ratio than the prototype, and physical conditions outside the boundary do not necessarily conform to the internal similarity criteria. Thus, there may arise at the boundary surface departures from similarity

105

which are termed *boundary*, or *wall*, *effects* and which may, unless controlled, render it almost impossible to predict large-scale performance from model experiments. An example would be a packed-tower element in which the tower diameter was only two or three times the ring diameter. Such an apparatus would give little information about the behavior of a full-scale tower owing to the predominant wall effect in the element.

Boundary effects cannot be eliminated by increasing the extent of the system under control. In the case of a model reaction vessel losing heat by natural convection to the surrounding air, an attempt might be made to secure external similarity by enclosing the vessel in a chamber supplied with air at a suitable temperature. The air surrounding the vessel then becomes a part of the system being studied, and the boundary effect is merely transferred to the walls of the air chamber. But although boundary effects cannot theoretically be eliminated, they can sometimes be neutralized or compensated for by suitable experimental devices.

The wall of a containing vessel may have four different kinds of effect upon a physical or chemical reaction proceeding within it:

1. It influences the fluid-flow pattern and frictional resistance.

2. It can transfer heat into or out of the system.

3. It may adsorb matter from or release matter to the fluid stream.

4. It may positively or negatively catalyze a chemical reaction in the fluid phase.

Flow Pattern

Under a viscosity-controlled dynamic regime, a geometrically similar model suffers no frictional wall effects. The difference in both interior and boundary surface/volume ratio is compensated by the increased velocity in the model. At equal Reynolds numbers, the fluid-flow patterns in model and prototype are similar.

In the case of an element, the wall effect becomes pronounced when the section is so far reduced that the boundary surface is of the same order of magnitude as the interior surface. This sets a practical limit to the section ratio that can be employed. (The section ratio is the ratio of cross section of the prototype to that of the element.) For example, it is found that the wall effect in a packed tower becomes serious when the tower diameter is less than about ten times the diameter of the packing. In multitubular construction, an element with too few tubes will have a hydraulic mean radius outside the tubes which is appreciably different from that in the prototype and which will give rise to different film conditions.

There is no simple method by which the effects on flow pattern and frictional drag of the higher surface/volume ratio of an element can be counteracted. In the case of packed towers with liquid descending and gas or vapor rising, it would appear advantageous to coat the inner wall of the element with some substance not wetted by the liquid, thus reducing the wall effect. In general, the best course is to limit the section ratio of an element to a value such that the ratio of boundary to interior surface does not exceed 10 per cent.

For evaluating frictional resistance, a scale model is preferable to an element or model element, provided that the corresponding velocity for similarity is not excessive, because in the model at equal Reynolds numbers there is no wall effect.

Heat Loss (or Gain)

Of the three main transfer processes involving, respectively, momentum, heat, and mass, heat transfer is the only one which penetrates the wall of the containing vessel. Hence, in operations and reactions which take place either above or below the ambient temperature, heat flow through the external surface of the apparatus is usually the principal wall effect to be guarded against.

Damköhler (7) based his treatment of thermal similarity upon the assumption that the surface-heat loss from a vessel is proportional to the internal-film coefficient. It has been pointed out (18) that in general this is not true. There is always an external-film coefficient which is outside the system and not influenced by internal-flow conditions. External-film coefficients due to natural convection tend to be of a lower order than internal-film coefficients where there is forced convection and turbulent flow. Further, where a process vessel is much above or below the ambient temperature, it is normally insulated for economic reasons. The net result is that in general the thermal resistance to heat transfer of the internal fluid film is negligible compared with the combined resistances of the external air film and insulation if any. Therefore, although internal-film coefficients vary with the fluid velocity, the over-all coefficients from the interior of the containing vessel to the surrounding air tend to be of the same order in both model and prototype, provided that where the vessels are insulated the same thermal resistance per unit area is applied to both (e.g., equal thicknesses of the same insulating material).

A second consideration is that, owing to the predominant effect of the external coefficient of heat transfer, the surface-heat losses from an experimental-process vessel can in practice be adjusted independently of temperature and flow conditions within. By jacketing the vessel with suitable heating or cooling media or by supplying electrical heat through a winding of resistance wire, the rate of heat loss can be controlled to any desired value down to zero. The jacket or winding applied to a hot vessel on the small scale is usually termed "adiabatic," although its true function is not to maintain truly adiabatic conditions but to reduce heat losses per unit of throughput to the same value as would obtain on the large scale. However, a full-sized plant efficiently insulated is so nearly adiabatic that it is generally sufficient to approximate to adiabatic conditions in the model.

There are three cases in which a small-scale apparatus can exhibit what may be called *inherent thermal similarity*, i.e., where the heat flux per unit external area is required to be the same in both prototype and model or element, so that no special jacketing of the latter is necessary. It is merely given the same thickness of insulation as the prototype. Two of these cases of inherent thermal similarity have already been noted in Chap. 7.

1. *Thermal regime.* Model heat-transfer systems in which the controlling mechanisms are radiation and conduction through the walls of the vessel [Eqs. (7-21)ff.].

2. *Chemical regime.* Model heterogeneous reaction systems with fixed interface in which the internal surfaces (catalyst grain size, etc.) are geometrically similar and in the same scale ratio as the reaction vessels, their surface activity remaining constant [Eqs. (7-29)ff.].

3. *Dynamic regime.* Inherent thermal similarity under a dynamic regime is impossible for either models or elements but can be secured by means of a model element of suitable proportions.

For similarity under a *dynamic regime*, the surface-heat loss per unit area in a scale model is required to be L times as great as that in the prototype, where L is the linear scale ratio of prototype to model. When operating above atmospheric temperature, such a model would need to be externally cooled, e.g., by water-jacketing. On the other hand, a dynamically similar element is required to lose less heat per unit of external area than the prototype. For vessels of elongated shape in which heat losses from the ends can be neglected, the surface-heat loss per unit area of the element should be $1/\sqrt{A}$ times that in the prototype, where A is the section ratio, i.e., the cross-section ratio of prototype to element. Evidently, there must be one particular set of model elements in which the conflicting requirements cancel out and which possess inherent thermal similarity. For elongated vessels, this is the case when

$$L = U\sqrt{A} \qquad (9\text{-}1)$$

where **U** is the ratio of over-all boundary surface-heat transfer coefficients for prototype/model element. Generally, **U** can be taken as unity, and the relation becomes

$$A = L^2 \qquad (9\text{-}2)$$

The same relation holds for a process operating at subatmospheric temperatures in which the wall effect consists of heat transfer inward.

The model-element relation is possible only where the prototype apparatus has a multiple or grid structure with a high ratio of interior to boundary surface. Plain vessels cannot possess inherent thermal similarity under a dynamic regime.

In all of the above cases of inherent thermal similarity, the reacting systems are considered, for similarity purposes, to be bounded by the internal surfaces of their containing vessels. The thickness of the vessel wall is not subject to the requirement of geometrical similarity and should, in conjunction with any external insulation, offer the same thermal resistance per unit area in both model (or model element) and prototype. In the case of a metallic vessel without insulation, the principal thermal resistance lies in the air film outside the vessel, and the effect of wall thickness can in general be neglected.

In systems for which inherent similarity is not possible, thermal conditions at the boundary must be artificially controlled. For the control of surface-heat losses from hot vessels, the electrical-resistance winding is usually the most convenient method. It is best embedded in a layer of thermal insulating material about halfway between the hot wall of the vessel and the surface of the insulation. Successive turns of wire should not be farther apart than the thickness of the layer of insulation between them and the vessel. This will give the total length of wire required. A generous estimate of the maximum heat loss from the winding through the outer layer of insulation (assuming the winding to be at the same temperature as the vessel), converted into electrical units, gives the maximum power input. From these figures and a knowledge of the resistivity of the metal, the required cross section of the resistance wire can be calculated.

Regulation of the heat input to the resistance winding is by a series rheostat or choke coil. For the maintenance of adiabatic conditions, it is sufficient to have one thermocouple inside the hot vessel and another embedded in the insulation between the wall and the resistance winding, the current being adjusted to keep the two temperatures equal. Where heat losses are to be controlled to a definite low value, it is better to have two sensitive thermocouples outside the vessel, one adjacent to the wall and the other close under the winding.   Knowing the thickness of insulation between them and the resistivity of the insulating material, one can accurately calculate the heat flow from the temperature difference.

Where the internal temperature varies along the length of a vessel, the resistance winding is commonly made in sections, each section being regulated by a separate rheostat and controlled by separate thermocouples.

Vacuum jackets are often used for thermal insulation in the laboratory, as in the ordinary Dewar flask, but they are troublesome and seldom effective on the pilot-plant scale.   Good insulation is not obtained unless the absolute pressure in the jacket is extremely low, under $\frac{1}{10}$ mm Hg, and it is difficult to maintain so high a vacuum in ordinary engineering equipment owing to leaking joints, porosity of metal and welds, and the vaporization of traces of moisture and oil. A vacuum jacket does not greatly diminish heat loss or gain by radiation, for which purpose highly reflecting metallic surfaces are necessary.

Surface Catalysis

The important thing about surface catalytic effects is that they should be detected on the small scale and allowed for in the design of large-scale plant.   Many apparently homogeneous chemical reactions are to some extent catalyzed by the walls of the reaction vessel and so would give different yields in a large vessel under similar time-temperature conditions owing to the lower ratio of surface to volume. The effect of providing extra surface in the reaction vessel can usually be investigated in the laboratory.   Where there is an appreciable surface catalytic effect, some form of packed reaction vessel affording a large interior surface would normally be used and such a vessel is readily scaled up in accordance with the above principles and those of Chap. 7.

A difficulty arises with simple vessels where it is geometrically impossible to provide as low a surface/volume ratio in the laboratory as on the large scale.   It may occasionally be feasible to coat the internal surface of a laboratory reaction vessel with some inactive substance in order to suppress surface catalysis.   For example, the reaction between ethylene and bromine is catalyzed by a glass surface but practically ceases when the glass is coated with paraffin wax.

The well-known effect of vessel dimensions on the size of particles produced by precipitation reactions is possibly akin to a surface catalytic effect.   Other conditions being equal, larger reaction vessels tend to give coarser and more readily filterable precipitates. This is one of the boundary effects which are often favorable to scaling up.

SYMBOLS IN CHAPTER 9

A = cross-sectional area
L = linear dimension
U = over-all heat-transfer coefficient

CHAPTER 10

DUCTS AND FLOW PASSAGES

The formulas of Poiseuille, D'Arcy, Fanning, or Chézy in conjunction with the Reynolds-Stanton curve of friction factor vs. Reynolds number and empirical allowances for bends, valves, etc., enable the frictional resistance of a normal piping or ducting system to be calculated with an accuracy sufficient for engineering purposes. Data are available for calculating the pressure drop through tower packings and beds of broken solids. Difficulty arises when a fluid is required to flow through passages of irregular shape with sudden changes of direction and cross-sectional area, constrictions and obstacles, divergence and recombination of channels, etc. Useful information can then be obtained from tests carried out with a geometrically similar model.

The fundamental equation for frictional pressure drop in an incompressible fluid flowing through a straight pipe or duct of constant cross section which it entirely fills may be written (100)

$$\frac{2g\,\Delta p}{\rho v^2} = 8\,\frac{r}{\rho v^2}\frac{l}{d} \tag{10-1a}$$

where Δp = frictional pressure drop, psf

g = 32.2 ft/sec²

ρ = fluid density, lb/ft³

v = mean fluid velocity, fps

r = frictional resistance per unit area of wetted surface, poundals/ft²

l = length of duct, ft

d = hydraulic diameter, ft = 4 times the cross-sectional area divided by the perimeter; in a cylindrical pipe this is equal to the geometrical diameter

It is customary to put $2r/\rho v^2 = f$, a friction factor, leading to the D'Arcy or Fanning equation in dimensionless form,

$$\frac{g\,\Delta p}{\rho v^2} = \frac{4fl}{2d} \tag{10-1b}$$

where f is a function of the Reynolds number and l/d is a ratio or shape factor defining geometrical similarity in straight channels of constant cross section. The most generalized form of the relation for geometrically similar channels of any shape is

$$\frac{g\,\Delta p}{\rho v^2} = \phi\left(\frac{\rho v d}{\mu}\right) = F \tag{10-1c}$$

where F is a function embodying both the friction factor f and the shape factor for the system. For straight pipes,

$$F = \frac{4fl}{2d}$$

It is convenient to be able to calculate Reynolds numbers from engineering units without converting them to consistent units, which can be done by the following formulas:

$$\text{Re} = \frac{\rho v d}{\mu}$$

$$= 7{,}725\,\frac{sv D}{\eta}$$

$$= 3{,}789\,\frac{sQ}{\eta D} \qquad \text{for cylindrical pipes only}$$

$$= 3{,}158\,\frac{sQ'}{\eta D} \qquad \text{for cylindrical pipes only}$$

where ρ = fluid density, consistent units (lb/ft³)
  s = specific gravity of fluid
  v = mean velocity, fps
  d = diameter or hydraulic diameter, ft
  D = diameter or hydraulic diameter, in.
  μ = viscosity, consistent units (lb/ft/sec)
  η = viscosity, centipoises
  Q = volumetric rate of flow, Imperial gpm
  Q' = volumetric rate of flow, U.S. gpm

Where the fluid is a gas and therefore compressible, Eq. (10-1e) is still a good approximation provided the total pressure drop is less than about 10 per cent of the total absolute pressure. When the pressure drop becomes larger in relation to the absolute pressure, this equation no longer applies because of the change in volume of the gas as its pressure falls. Flow conditions are then represented by the differential equation

$$\frac{g\,dp}{\rho v^2} = \frac{4f\,dl}{2\,d} \tag{10-2a}$$

For isothermal flow of any gas in a tube, this may be approximately integrated as follows:

Conservations of mass along the tube give

$$\rho_1 v_1 = \rho v = \rho_m v_m = \rho_2 v_2$$

where ρ and v represent the density and mean linear velocity of the gas and the subscripts 1, 2, and m refer, respectively, to conditions at the inlet, outlet and the arithmetic mean of these.

The equation of state of the gas is

$$\rho = \rho_1 \frac{z_1}{z} \frac{p}{p_1} \frac{T_1}{T}$$

where z is the correction factor for the deviation from the perfect-gas law, i.e., the compressibility factor. For isothermal flow, $T_1 = T$. Thus, Eq. (10-2a) becomes

$$\frac{g \, dp \, \rho}{\rho_1{}^2 v_1{}^2} = \frac{4F \, dl}{2d}$$

or

$$\frac{g \, dp \, \rho_1}{\rho_1{}^2 v_1{}^2} \frac{z_1}{z} \frac{p}{p_1} = \frac{4f \, dl}{2d} \qquad (10\text{-}2b)$$

If we replace the variable z by z_m, which is a nearer approximation than taking the value of z at either end, Eq. (10-2b) can be integrated to

$$\frac{g(p_1{}^2 - p_2{}^2)}{p_1 \rho_1 v_1{}^2} \frac{z_1}{z_m} = \frac{2fl}{2d} \qquad (10\text{-}2c)$$

If we define $p_m = (p_1 + p_2)/2$, then this is equivalent to

$$\frac{g \, \Delta p}{\rho_1 v_1{}^2} \frac{p_m}{p_1} \frac{z_1}{z_m} = \frac{fl}{d} \qquad (10\text{-}2d)$$

where

$$\Delta p = p_1 - p_2 \qquad (10\text{-}3)$$

In the case of a perfect gas, the term z_1/z_m vanishes.

Compressibility factors for common gases are published in standard works (101, 141). Alternatively, if the critical constants of the gas are known or can be estimated, its compressibility factor may be closely approximated from generalized curves such as those of Obert (111). For more accurate readings, reference may be made to the large-scale charts published later by Nelson and Obert (110).

For gaseous-flow systems in which the pressure loss is high in relation to the absolute pressure of the system, dynamic similarity is not possible when the model also is operated with the gas. Equality of pressure drop in model and prototype is inconsistent with equality of Reynolds number. In such cases, the best procedure is to calibrate the model with water over the required range of Reynolds numbers and determine F in Eq. (10-1c). The appropriate value is then substituted in Eqs. (10-2c) or (10-2d).

Equations (10-2b) to (10-2d) can be solved directly for known end conditions, but to find the end conditions for a known rate of flow, a trial-and-error solution is necessary. F is found for a given Reynolds number by calculation or experiment. v_1, p_1, and ρ_1 are fixed. A value of p_2 is assumed to give p_{av}, from which Δp is calculated. A more accurate value of p_2 can then be taken and the calculation repeated.

The geometrical similarity of the model should extend at least approximately to surface roughness. That is, the surfaces in contact with the moving fluid should be proportionately smoother in the model than in the prototype. The percentage error caused by neglecting surface roughness increases with the Reynolds number in a manner illustrated by the following table, which shows the approximate ratio of friction factors for commercial iron pipes and glass tubes at equal Reynolds numbers:

Reynolds number	Ratio of friction factors, $\dfrac{\text{rough pipes}}{\text{smooth tubes}}$
10^3 (streamline)	1.00
10^4	1.15
10^5	1.25
10^6	1.65

Where homologous systems (i.e., the same fluids) are employed on both the small and the large scale, the requirement of equal Reynolds numbers limits the scale ratios that are practicable. In gaseous systems, corresponding velocities in small-scale models may be so high that compression effects become appreciable, and the Reynolds number is no longer the sole criterion of similarity. Even with liquids, where compression effects are negligible, an excessive power consumption is required to maintain in a small scale model velocities corresponding to normal practice on the large scale. The classical method of overcoming this difficulty is to test the model, using a fluid of much lower kinematic viscosity than that to be employed on the full scale. The lower the kinematic viscosity in the model, the lower the fluid velocity corresponding to a given velocity in the prototype. Figure 10-1 shows kinematic viscosity μ/ρ plotted against temperature for hydrogen, air, water, and mercury. At 20°C, the kinematic viscosity of hydrogen is about $6\frac{1}{2}$ times that of air, 100 times that of water, and 900 times that of mercury.

When the Reynolds number in a flow system exceeds about 10,000 and the system consists mainly of rough pipes, bends, and irregular changes of section rather than of long, smooth channels, the flow pattern and friction factor f become substantially independent of Reynolds number. In these circumstances, a model can often give accurate information even though it is operated at a Reynolds number

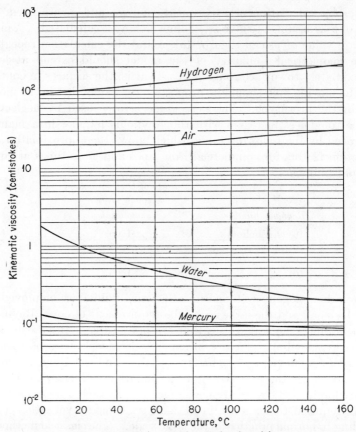

Fig. 10-1. Range of kinematic viscosities.

very much below that of the prototype, provided always that the value in the model exceeds 10,000. This device is often used in cold-flow models of furnaces (see Chap. 16).

A typical experimental setup for carrying out frictional-resistance tests with liquids is shown in Fig. 10-2. It comprises the model, circulating pump, control valve, flowmeter, differential manometer, tank, and thermometer. If possible, the liquid circuit should be laid out with a continuous rise to the outlet and no air pockets. If owing to the shape of the model air pockets are unavoidable, means must be provided for bleeding the air out of them so that the system is completely filled with water.   Before the liquid inlet to the model, there should be a calming section of pipe at least 20 diameters in length to smooth out disturbances caused by bends and fittings.   Straightening vanes in the section help to smooth the flow but are not essential unless the calming section is below 20 diameters in length.   The circulating pump should be of the centrifugal or positive rotary type, giving a flow that is free from pulsation.   A thermometer in the liquid is necessary in order that its density and viscosity may be accurately known.

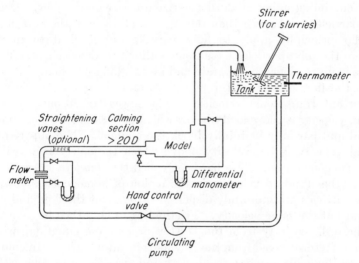

FIG. 10-2. Apparatus for fluid-friction experiments, with a water model.

Models can be used to determine the coefficients of discharge of metering orifices when these are of such a shape or have to be installed in such a position that the standard orifice formulas do not apply. Here it is necessary to achieve equality of Reynolds number and fairly exact geometrical similarity of surface roughness.   Any build-up of dust or sediment on the large scale must also be reproduced in the model.   Under these conditions, water or cold air may be used in a model to give a Reynolds number equal to that of the hot gases in a large furnace, thus greatly reducing the corresponding velocity in the model, as mentioned above.   Chesters and Thring (102) employed this technique to calibrate a short venturi orifice constructed in refractory concrete and used to meter hot raw producer gas to an open-hearth furnace.

Slurry Pumping

It is seldom possible to predict with any confidence the flow characteristics and frictional losses of solid liquid slurries, even in straight pipes. The empirical formula of Durand, which is given below, applies only to closely graded particles suspended in water. For normal industrial solids having a wide range of particle size, and for slurrying liquids other than water, it is usually advisable to carry out experiments with the particular slurry to be handled before designing a slurry-pumping system.

A slurry is a liquid suspension of solid particles which tend to separate out on standing or during horizontal streamline flow. Usually the particles are heavier than the liquid and settle to the bottom, but similar principles apply to light particles which float to the top. Where the solid particles have exactly the same density as the liquid, there is no tendency to separate and no handling problem.

In the flow of slurries through pipes and channels, there are two important transitional velocities. The lower transitional, or minimum, velocity is the velocity below which particles settle out from the liquid and partially or totally block the channel. The upper transitional velocity is the velocity above which the slurry behaves as a homogeneous liquid. This has been called the *standard velocity* (115). Above the standard velocity, in the region of homogeneous flow, the solid particles are uniformly dispersed throughout the fluid and move forward at the same velocity. Friction losses may then be calculated in the ordinary way from the D'Arcy or Fanning equation, using a friction factor derived from the Reynolds-Stanton curve. In computing the Reynolds number, the density and viscosity are taken as those of the slurry.

Between the minimum velocity and the standard velocity lies the region of heterogeneous flow. In this region, the solids advance more slowly than the liquid, and in a horizontal pipe there is an appreciable concentration gradient from top to bottom. Friction losses are considerably higher than for the pure liquid at the same velocity. The added frictional resistance may be visualized as partly due to the fluid having to percolate forward through the suspension of more slowly moving solids.

The major factors in determining the minimum and standard velocities for a particular slurry are the size of the particles and their apparent density (or buoyancy) when immersed in the fluid. For very fine, or low-density, slurries, the standard velocity may be below the range of economic pumping velocities. Within that range, the slurry behaves as a liquid, and there is once again no handling problem. For many of the coarser industrial slurries, such as sand or broken coal suspended in water, the standard velocity is far above the range of economic pumping velocities. Such slurries must be handled in the region of heterogeneous flow, and it is there that experiments are particularly necessary.

In designing a slurry-pumping system, there are two main problems: first, to determine the minimum and standard velocities in order to establish whether flow will be in the homogeneous or the heterogeneous region; second, to calculate the frictional resistance to flow at the selected velocity.

Spells (115) applied dimensional analysis to the problem of minimum and standard velocities and showed that both depend essentially upon a balance between the dispersing effect of the turbulence and the gravitational forces tending to concentrate the particles at the bottom of the pipe. Hence, the Froude number, representing the ratio of inertial to gravitational forces, will be related to the Reynolds number, representing the ratio of inertial to viscous forces. The inertia of the moving liquid depends upon its absolute density, whereas the gravitational force acting on a solid particle depends upon its absolute density immersed in the liquid. Hence, the Froude number needs to be modified to include the ratio of these densities. Spells concluded that both minimum and standard velocities should conform to a dimensionless equation of the type

$$\frac{\rho v^2}{\sigma g \delta} = \phi\left(\frac{\rho' v d}{\mu}\right) \tag{10-4}$$

where v = either minimum or standard velocity
d = pipe diameter
g = acceleration of gravity
δ = mean particle diameter
ρ = density of liquid
ρ' = mean density of slurry
σ = "effective density" of immersed particles = $\rho_{\text{solid}} - \rho$
μ = viscosity of liquid

For particles between 50 and 500 μ in diameter flowing in straight horizontal pipes, he proposed the empirical equation

$$\frac{\rho v^2}{\sigma g \delta} = K\left(\frac{\rho' v d}{\mu}\right)^{0.775} \tag{10-5}$$

where $K = 0.0251$ for the minimum velocity, 0.0741 for the standard velocity.

Correlation with published experimental results is only approximate, and Spells emphasizes that Eq. (10-5) should not be used outside the particle-size range mentioned. It does, however, offer a basis for scaling up experimental determinations of minimum and standard velocity on the small scale. For homologous systems having the same range of particle size, Eq. (10-5) yields the scale-up relation

$$\mathbf{v} = (\varrho'd)^{0.63} \tag{10-5a}$$

where $\rho' = C\sigma + \rho$ and C = volumetric concentration of solids.

The second problem is to scale up the frictional resistance. Fundamental differential equations corresponding to the Navier-Stokes equations for liquids have not been worked out for liquid solid slurries. The generalized dimensionless equation must therefore be derived from dimensional analysis plus some knowledge of the physical mechanisms at work. The first mechanism that is not present in the flow of pure liquids is the settling of the solid particles. The law of settling of particles in a quiescent liquid depends upon the Reynolds number $\rho u \delta / \mu$, where u is the settling velocity and δ is the particle diameter. For Re < 2, the settling velocity follows Stokes' law, corresponding to streamline flow of liquid past the particle. For Re > 500, the settling velocity follows Newton's law, corresponding to fully developed turbulence in the wake of the particle.

Stokes' law of settling is

$$u = \frac{g\delta^2\sigma}{K\mu}$$

where u = settling velocity relative to the liquid and K = a const.

Newton's law of settling is

$$u = \sqrt{K'\frac{g\delta\sigma}{\rho}}$$

The generalized dimensionless equation for slurry flow may be expected to contain either the group $g\delta^2\sigma/\mu u \equiv K$ or the group $\rho u^2/g\delta\sigma \equiv K'$ according as Stokes' or Newton's law applies. Both K and K' are shape factors depending only on the geometry of the solid particles.

The other physical process at work is the lifting of particles into the fluid stream against the force of gravity through the transfer of momentum from liquid to solids. The physical variables involved are the mean fluid velocity v, fluid density ρ, effective particle density

$$\sigma = \rho_{\text{solid}} - \rho$$

pipe diameter d, and acceleration of gravity g. Those give the dimensionless group $\rho v^2/g\, d\sigma$, representing the ratio of inertial forces in the liquid to gravitational forces acting on the solids. The final variable is the concentration of the slurry, which may be expressed as C, the volumetric fraction of solids. This is in itself a dimensionless ratio. If $\Delta h'$ denotes the loss of head per unit length of pipe or channel due to friction, the generalized dimensionless equation may be written

$$\frac{\Delta h'}{v^2} = \phi\left(C, \frac{\rho'vd}{\mu}, \frac{\rho v^2}{gd\sigma}, K\right) \qquad (10\text{-}6a)$$

if Stokes law applies, or

$$\frac{h'}{v^2} = \phi\left(C, \frac{\rho'vd}{\mu}, \frac{\rho v^2}{gd\sigma}, K'\right) \qquad (10\text{-}6b)$$

if Newton's law applies.

It is convenient to represent the friction loss in the flow of a slurry as a ratio of the loss at the same velocity with the pure liquid, denoted by Δh. This ratio, within limits, allows the results of slurry-pumping experiments to be extrapolated to systems of different geometrical form. Since $\Delta h = \phi(\rho vd/\mu)$ and $\rho'/\rho = \phi C$, we may write either

$$\frac{\Delta h'}{\Delta h} = \phi\left(C, \frac{\rho v^2}{gd\sigma}, K\right) \qquad (10\text{-}7a)$$

or

$$\frac{\Delta h'}{\Delta h} = \phi\left(C, \frac{\rho v^2}{gd\sigma}, K'\right) \qquad (10\text{-}7b)$$

The most comprehensive experiments on the flow of slurries are those of Durand and his coworkers (103), who correlated their results in terms of a function $\varphi = (\Delta h' - \Delta h)/\Delta h$. For aqueous slurries, they proposed an empirical equation which may be written

$$\varphi = 121C\left(\frac{gd\sigma}{v^2}\right)^{1.5}\left(\frac{u^2}{g\delta\sigma}\right)^{0.75} \qquad (10\text{-}8)$$

The term σ was introduced into the original Durand equation by Worster (118) to correlate results with particles of different densities.

This has the same general form as Eq. (10-7b) since where ρ (in metric units) $= 1$, the group $u^2/g\delta\sigma \equiv K'$. In other words, Durand's results are correlated on the assumption that Newton's law of settling applies throughout, even though his experiments covered fine slurries, the particles of which should have obeyed Stokes' law. This may be due either to the state of turbulence already existing in the liquid or to the fact that where particles are fine enough to obey Stokes' law the standard velocity is near or even below the customary pumping velocities of 5 to 6 fps so that the flow of such slurries is substantially homogeneous. The interesting point about Eq. (10-7b) and (10-8) is that the pressure drop in slurry pumping depends not on particle size but only on shape, density, and volumetric fraction of solids in the slurry.

From the definition of φ,

$$\frac{\Delta h'}{\Delta h} = \varphi + 1$$

The Durand-Worster equation (10-8) was derived from experiments with closely graded solid particles, and it may not be directly applicable to ordinary industrial slurries having a wide range of particle sizes. It may, however, be taken as a basis for scaling up experiments with a given type of slurry. Assuming the same particle shape, size spectrum, and effective density on the small and large scale, Eq. (10-8) may be written

$$\frac{\boldsymbol{\varphi}}{\mathbf{C}} = \left(\frac{\mathbf{d}}{\mathbf{v}^2}\right)^{1.5} \tag{10-8a}$$

boldface type, as usual, denoting ratios of corresponding quantities on the large/small scale. Equation (10-8a) should apply even though the mean particle sizes on the large and small scale are different, provided only that the curves of size distribution about the mean are similar.

The procedure for scaling up the flow characteristics of a particular slurry is to set up an apparatus similar to that shown in Fig. 10-2. The "model" in the case may be a length of pipe, preferably transparent so that settling of the solids can be observed. The minimum velocity is determined visually and the pressure drop measured over a range of velocities for both the slurry and the pure liquid. φ/C is then plotted against d/v^2. The observed minimum velocity may be scaled up in accordance with Eq. (10-5a).

In the design of the experimental apparatus, it is important that the velocity in every other part of the fluid path (except the stirred tank) should be higher than that in the test section so that settling starts there before it takes place anywhere else.

SYMBOLS IN CHAPTER 10

C = fractional volumetric concentration of solids in a slurry
D = diameter of pipe, in.
d = diameter of pipe, ft, or hydraulic diameter of a noncircular channel, 4 times the cross-sectional area divided by the wetted perimeter

f = friction factor in the D'Arcy or Fanning equation

g = acceleration of gravity, ft/sec²

Δh = friction head per unit length of pipe in a pure liquid flowing at velocity v

$\Delta h'$ = friction head per unit length of pipe in a slurry flowing at velocity v

K, K' = constants

l = length of pipe or duct, ft

p = pressure, psf

Δp = frictional pressure drop, psf

Q = volumetric rate of flow, Imperial gpm

Q' = volumetric rate of flow, U.S. gpm

r = frictional resistance per unit area of wetted surface, poundals/ft²

s = specific gravity

u = settling rate of a solid particle of diameter δ

V = specific volume of a gas = $1/\rho$

v = mean fluid or slurry velocity, fps

z = compressibility factor of a gas = pv/RT

δ = mean particle diameter in a slurry

η = viscosity of fluid, centipoises

μ = viscosity of fluid, fps

ρ = fluid density, lb/ft³

ρ' = mean density of a slurry = $C\sigma + \rho$

σ = "effective density" of solid particles immersed in a liquid = $\rho_{\text{solid}} - \rho$

ϕ = a function of $\cdots$

φ = $(\Delta h' - \Delta h)/\Delta h$

Subscripts

1, 2 = relating to entry, exit of pipe or channel

m = arithmetical mean value

CHAPTER 11

FILTERS

Filtration is a mechanical separation of solids from fluids and is therefore subject to a dynamic regime.   In practice, model theory does not enter into the scaling up of filtration experiments because the absolute magnitudes of the solid particles and fluid passages, as well as their geometrical shapes, are determined by the conditions of the process.   The only direct method of scaling up experimental results is to have a small-scale filter that corresponds to an element of the full-sized unit.

Theoretically, there would seem to be no difficulty in scaling up the performance of an experimental filter element in direct proportion to the area or of extrapolating the experimental results to higher pressures or thicker filter cakes by means of the empirical rate equations to be discussed below.   In practice, large-scale filtration rates are apt to differ substantially from those predicted on the basis of small-scale experiments, and it is advisable to allow an ample margin of safety in specifying the final filter area.   One of the principal difficulties is not the filtration itself but the preparation on a small scale of a slurry that shall have the same filtering properties as that obtained on the large scale.   In many precipitation reactions, there is a marked scale effect due possibly to differences in degree of turbulence, rate of cooling, or rate of nucleation either by adventitious foreign matter or by the walls of the containing vessel, heat-transfer surfaces, etc.   The result is that the small-scale precipitate may have a different particle size and porosity from that obtained on the large scale.   Such scale effects can be reduced by choosing appropriate experimental conditions (see Chaps. 12, 13, and 15), but it is better if possible to carry out filtration experiments using samples of slurry drawn from a large-scale plant.   Even when slurry differences are eliminated, the experiments of McMillen and Webber (94) show that considerable discrepancies are possible between predicted and actual filtration rates on the large scale, the actual rates being usually lower than those predicted.

Because of the small hydraulic diameter of the liquid passages, the flow of filtrate through a filter bed is streamline and conforms to a modified Poiseuille's or D'Arcy's law. The instantaneous rate of flow is directly proportional to the pressure and inversely proportional to a resistance which is the sum of two terms, the resistance of the filter cake and that of the filter septum (cloth, gauze, etc.). The former resistance increases with the thickness of the cake or the volume of filtrate that has passed; the latter is a constant. On the assumption that the filtration pressure remains constant, the rate equation may be written

$$\frac{dq}{dt} = \frac{p}{aq + b} \tag{11-1}$$

where q = total volume of filtrate passing through *unit area* of filter bed in time t

p = total pressure drop across filter cake and filter septum

a = a const related to the viscosity of the filtrate and the physical properties of the filter cake

b = a const related to the viscosity of the filtrate and the physical properties of the septum

This is the familiar Ohm's-law type of rate equation in the series form. Inverting and integrating between time limits 0 to t, and volume limits 0 to q_1,

$$t_1 = \frac{\frac{1}{2}aq_1{}^2 + bq_1 + c}{p} \tag{11-2}$$

The constant c is equal to t_1 when $q_1 = 0$ and may be considered to represent the time interval after filtration starts and before any filtrate appears, i.e., the time required to fill the filtrate compartment and pipes, etc.

Ruth (97) gives the integrated constant-pressure filtration equation in a rather neater form,

$$(q_1 + q_0)^2 = K(t_1 + t_0) \tag{11-3}$$

where K = a const which is directly proportional to p

q_0, t_0 = constants related to the resistance of the septum

q_0 can be interpreted as the volume of filtrate that would be required to deposit a filter cake of resistance equal to the resistance of the septum and t_0 the time that would be taken to form such a cake. Ruth's equation contains no separate constant for the volume of filter pipes.

The foregoing equations assume that the resistance to filtration is independent of the pressure, i.e., that both the filter cake and the filter septum are incompressible. In fact, all filter cakes and septa are compressible to some extent. Lewis and his coworkers (92, 143) made the empirical assumption that the resistance of the filter cake varied as p^s and that of the filter septum as p^m. Applying these corrections to Eq. (10-1) and inverting it,

$$\frac{dt}{dq} = \frac{a'q}{p^{1-s}} + \frac{b'}{p^{1-m}} \qquad (11\text{-}4)$$

where a' and b' are modified constants.

In most industrial filtrations, the value of s lies between 0.1 and 0.8, while m tends to be somewhat lower.

The assumption of a simple power relation between resistance and pressure is an approximation which holds only over a limiting range of pressures. For example, the cake resistance does not approach zero as the pressure approaches zero. Carman (93) has suggested that the relation would be better represented by a function of the type $1 + \alpha p^s$, where α is a constant.

Filtration experiments are generally conducted batchwise and at constant pressure, since the results can then be readily correlated on the basis of a simple constant-pressure equation such as (11-3) or (11-4). The general method is to take periodic observations of the filtration time t and total weight or volume of filtrate $Q(= qA)$ until the required thickness of filter cake has been built up. Where the filter cake is compressible and the experimental results are to be extrapolated, the batch experiment should be repeated at two or three different pressures and with other conditions the same.

The experimental results may be correlated graphically by either of two methods, one based on the integral form of the constant-pressure equation and the other on the differential form. The first method was originally suggested by Underwood (99). If the total volume of filtrate per unit area q_1 is corrected for the volume of the filtrate compartment and pipes so that the constant of integration disappears, Eq. (11-4) on integration becomes

$$t_1 = \frac{a'}{2p^{1-s}} q_1{}^2 + \frac{b'}{p^{1-m}} q_1$$

or

$$\frac{t_1}{q_1} = \frac{a'}{2p^{1-s}} q + \frac{b'}{p^{1-m}} \qquad (11\text{-}5)$$

By plotting corresponding values of t/q versus q, a straight line should be obtained, the slope of which gives the value of $a'/2p^{1-s}$, while the intercept at $q = 0$ gives b'/p^{1-m}. After two or more constant runs have been carried out at different pressures, log-log plots of $a'/2p^{1-s}$ and b'/p^{1-m} versus p will give the values of s and m, respectively.

The second method of correlation, based on the differential form of the filtration equation, is used by Ruth and his coworkers (95, 96, 97). Experimental values of t are first plotted against q, and from this parabolic time-volume discharge curve instantaneous values of dt/dq are read off corresponding to various values of q. Differentiating Ruth's equation (11-3),

$$\frac{dt}{dq} = \frac{2}{K} (q + q_0) \qquad (11\text{-}6)$$

When dt/dq is plotted against q, a straight line is again obtained, the slope of which gives $2/K$ and the intercept $2q_0/K$. With compressible sludges, it is possibly better to work with Eq. (10-4), since this makes allowance for the different effects of pressure on the filter cake and septum. The slope of the plot of dt/dq versus q then gives a'/p^{1-s}, while the intercept at $q = 0$ gives b'/p^{1-m}. McMillen and Webber (94) have shown that with Ruth's method it is not necessary first to plot the time/volume discharge curve. By noting the increasing time intervals Δt required for equal increments of filtrate weight or volume ΔQ and plotting $\Delta t/\Delta Q$ versus Q (or $\Delta t/\Delta q$ versus q), a direct straight-line plot is obtained corresponding to Eqs. (11-4) or (11-6).

In the case of easily filterable and relatively incompressible materials, the design or selection of full-sized filters is sometimes based on laboratory experiments with simple vacuum filter leaves. For compressible cakes, a small pressure filter may be used such as that described by Sperry (98), in which the pressure is applied by compressed air. For experiments on a larger scale, laboratory filter presses are available in which full-scale conditions can be more closely duplicated, including conditions of variable pressure. The frames or plate recesses of the experimental press should preferably have the same depth as those of a full-sized press in order that the small apparatus may constitute a true element of the large one.

The term for septum resistance in the filtration equations corresponds to a filter cloth, or septum, that is (1) already impregnated with the finer particles of the solid and (2) stressed and deformed through supporting the filter cake against the pressure on the slurry side. Thus, the effective septum resistance in the actual filtration bears little relation to that measured by passing clear filtrate through the clean filter cloth alone. Where septum resistance constitutes an appreciable proportion of the total, the small-scale experiments should be carried out with a septum which has been used several times before. The resistance of a filter cloth that has been thoroughly impregnated with solids from previous filtrations may be several times that of a new cloth.

Ruth (96) has suggested that the data obtained from constant-pressure batch-filtration experiments can be applied to the design or selection of continuous rotary vacuum filters. So far as any given element of filter surface is concerned, continuous filtration is a succession of batch filtrations over the periods during which the filter element is immersed. For these conditions, Ruth derives a modification of his equation (11-3), which may be written

$$(q^1 + q_0)^2 = K(t^1 + t_0) \tag{11-7}$$

Where q^1 = filtrate volume per unit area of filter surface *per revolution*
t^1 = time of immersion of unit area of filter *per revolution*
K, q_0, t_0 = filtration constants identical with those for batch filtration

Ruth compared batch-filtration rates in a laboratory plate and frame press with those in a small Oliver filter using a precipitated calcium carbonate slurry, the effects of pressure being corrected by a plot of K versus p. The specific resistance of the filter cake on the Oliver filter was found to be about 90 per cent of that calculated from the filter-press results.

McMillen and Webber (94) conducted an extensive study of the results of scaling up filtration experiments which is much less favorable to extrapolation from laboratory press to continuous filter and which illustrates the difficulty of predicting large-scale filtration rates with accuracy from small-scale experiments even when variations in slurry properties are eliminated. Tests were carried out using the same slurry in a 6 by 6-in. laboratory press, a 12 by 12-in. industrial press, a 12 by 12-in. rotary vacuum-drum filter, and a 4-ft-diameter rotary vacuum-leaf filter. The filtration rates observed in the three larger filters were compared with those predicted from the laboratory-press results using Ruth's equation (11-3). Eighty-three direct comparisons were made and tabulated.

McMillen and Webber did not subject their results to a statistical examination, and their conclusions were perhaps more optimistic than the figures warranted. Table 11-1 summarizes a statistical analysis of their experiments.

Bearing in mind that the 95 per cent confidence limits are approximately equal to twice the standard deviation, one may draw the following conclusions:

1. Both types of rotary vacuum filter when filtering thin slurries (5 to 20 per cent of solids) gave rates of filtration averaging about half those predicted from the laboratory data. In the worst case, that of the 4-ft leaf filter, individual rates could be as low as 25 per cent of those predicted.

2. In filtering thick slurries (40 to 50 per cent of solids), the average filtration rates for the vacuum drum were about the same as those predicted, for the vacuum leaf about 25 per cent lower. Results were still very erratic, and ratios of observed to predicted rates in individual experiments might be half the average ratio or less.

3. Even when scaling up to another filter press of only three times the area, observed filtration rates could be occasionally as low as 66 per cent of those predicted.

TABLE 11-1. STATISTICAL ANALYSIS OF McMILLEN AND WEBBER'S
PREDICTIONS OF LARGE-SCALE FILTRATION RATES FROM
EXPERIMENTS WITH A LABORATORY PRESS

Type of large-scale filter	Scale-up factor (area basis)	Number of experiments	Range of slurry concentrations, wt %	Observed filtration rate as percentage of predicted rate	
				Average	Standard deviation
Filter press..............	3.16	16	2–15	93.2	13.4
Rotary vacuum drum.....	3.2	17	36–54	105.0	24.2
	3.2	16	9–26	60.7	21.9
Rotary vacuum leaf.......	18.0	17	36–53	74.6	19.5
	18.0	17	7–21	47.9	16.3

There is little doubt that the specialist manufacturers of filtration plant, using experimental equipment designed and calibrated to correspond to their own standard units, could do somewhat better than this. Nevertheless, in view of the very moderate scale-up factors in the McMillen and Webber experiments and the evident care with which the experiments themselves were conducted, the above analysis serves to underline the caution that is necessary in attempting to scale up this apparently simple operation. A factor of safety of 30 to 50 per cent would not be excessive. In particular, it appears inadvisable to extrapolate from batch experiments to continuous filters. Continuous-filtration constants are best determined on a continuous filter.

SYMBOLS IN CHAPTER 11

a, b = filtration const for incompressible cakes
a', b' = filtration const for compressible cakes
K = "const" in Ruth's equation (11-3), directly proportional to p

m = exponent on p for the filter septum
p = total pressure drop across filter cake and septum
q = total volume of filtrate passing unit area of filter bed in time t
q_0 = const in Ruth's equation (11-3)
s = exponent on p for a compressible filter cake
t = time
$_0$ = const in Ruth's equation (11-3)

CHAPTER 12

HEAT-TRANSFER EQUIPMENT

The types of equipment dealt with in this chapter are those in which heat is transferred from one fluid to another through a metal wall by conduction and convection: e.g., heat exchangers, steam- or vapor-heaters, evaporators, condensers, and coolers. Usually the metal wall takes the form of a tube or bundle of tubes, but there are also heat exchangers composed of flat plates or of strips coiled into a double spiral. In this class are included also regenerative heat exchangers such as those employed in liquid-air plants, where a cold gas first abstracts heat from a metal assembly and by means of a cyclic reversal of flows the cold metal then cools the incoming air. In the design of equipment of the above types, the effect of radiation is assumed to be negligible. Devices which depend upon heat transfer by radiation are treated in Chap. 16, Furnaces and Kilns.

In homologous systems under otherwise comparable conditions heat-transfer coefficients for conduction and convection tend to be higher on the small scale than on the large scale. Hence, if coefficients determined experimentally in a pilot plant are applied without adjustment to the full-scale design, the heat-transfer surfaces of the large plant are liable to be underdesigned. This chapter describes methods of adjusting or extrapolating coefficients determined on the small scale.

Industrial heat-transfer equipment is generally designed by calculation from the physical properties of the fluids rather than on the basis of model experiments. The equations used are partly theoretical and partly empirical. McAdams (161) states that his recommended equations for calculating heat transfer by forced convection inside and outside tubes reproduce experimental results with an *average* deviation of ±20 per cent.* This is for the simplest case without the interference of natural convection, boiling, or condensation. Where these occur, the deviations from calculated rates are likely to be

* "Average deviation" is unfortunately a mathematically intractable statistic. If the standard deviation had been given, some quantitative conclusions could have been drawn about errors, confidence limits, and factors of safety.

greater.   Among the causes of deviation are effects of shape on the fluid-flow pattern (including maldistribution in multitubular units), differences between the assumed average values of physical properties of the fluids and their actual point values throughout the apparatus, and traces of impurities occurring only on the large scale or after prolonged operation.   For example, fouling of a liquid-cooled surface can readily reduce the heat-transfer coefficient from liquid to metal to a quarter of its original value.   On the other hand, the presence of a trace of oil vapor in steam may increase the steam side film coefficient in a condenser six or eight times owing to the change from film to dropwise condensation.

Heat-transfer equipment is seldom the most costly item in a process plant, and its cost per square foot decreases as the size of the unit is increased.   Consequently, it is often less expensive to design units on paper with ample factors of safety than to carry out model experiments with the sole object of obtaining more accurate design data.   Where, however, a pilot plant must in any case embody heat-transfer equipment, it is worthwhile to design the units so that by the application of model theory their performance can be scaled up directly.   By this means, deviations due to the geometry of the system and the changing physical properties of the fluids are partly, if not entirely, eliminated. Manufacturers of heat-transfer equipment also may wish to have experimental apparatus which will enable them to predict the performance of their standard units under unfamiliar conditions.

If pilot-scale heat-transfer equipment is to provide useful performance data for larger units, it should preferably be designed for that purpose.   Where a preliminary engineering study of the process has been carried out (as recommended in Chap. 2), it will have indicated the types of heat exchanger envisaged on the large scale.   Corresponding types should be selected for those units in the pilot plant which it is desired to scale up.   The principal types of heat-transfer equipment used industrially are:

1. Coils immersed in tanks or stirred pots
2. Coils or serpentine pipes in air, dry or wetted externally with water
3. Double-pipe heat exchangers, plain or finned
4. Multitubular shell units with various numbers and combinations of passes
5. Plate or helical-strip heat exchangers

In a heat-transfer unit there are three successive processes at work: convection from the hot fluid to the metal wall, conduction through the wall and any adherent scale, and convection again from the wall to the cold fluid. The rate equation therefore takes the series form, the total thermal resistance being the sum of three or more resistances due to the two fluid films, the metal wall, and any scale or fouling which may be present on either side of the wall.

The similarity criteria for heat transfer between fluids depend upon the mechanism: whether transfer is by forced or natural convection, and whether the fluid systems are single-phase or two-phase with condensation or boiling. In natural-convection and two-phase systems, the heat-transfer process is complicated by gravity or buoyancy forces. Often the similarity criteria for the two sides of a heat-transfer unit are incompatible, as when there is forced convection on one side and natural convection on the other. In any case, it is seldom feasible to establish true similarity between model and prototype except when the scale ratio (or difference in size) is small. Usually the heat-transfer coefficient determined in the model must be extrapolated to kinematically dissimilar conditions on the large scale by means of an empirically determined exponent such as the Reynolds index. Sometimes the fluid-film resistance on one side is negligible compared with that on the other. Liquid-film coefficients are generally of the order of ten times as great as gas-film coefficients and boiling-liquid or condensing-vapor coefficients about ten times as great as liquid-film coefficients. If, therefore, a heat exchanger has liquid on one side and gas on the other, variation in the liquid-film coefficient will have a relatively small effect on the over-all coefficient and can be calculated for both model and prototype, leaving the gas side coefficients to be scaled up directly.

Heat transfer in process plants often takes place simultaneously with some other physical or chemical process having its own conditions of similarity. For example, the scaling up of a chemical reaction normally calls for equal mean residence times and, in the case of continuous reactors, a similar statistical distribution of residence times [i.e., Danckwerts's "F diagram" (189) for model and prototype should be identical]. Such process requirements may necessitate the extrapolation of pilot-plant results to give equal heat-transfer rates either per unit area or per unit volume on the large scale.

The similarity criteria and extrapolation relations for each side of a heat-transfer unit will first be treated separately and then methods given for scaling up the over-all heat-transfer coefficient.

Forced Convection

Heat transfer to a single fluid phase by forced convection conforms to the generalized dimensionless equations (7-12) and (7-14),

$$\frac{hL}{k} = \phi\left(\frac{\rho vL}{\mu}, \frac{c_p\mu}{k}\right) \tag{12-1a}$$

or
$$\frac{Q}{kL\,\Delta T} = \phi\left(\frac{\rho vL}{\mu}, \frac{c_p\mu}{k}\right) \tag{12-1b}$$

where L = film coefficient of heat transfer

Q = total heat transferred in unit time

L = characteristic length (or diameter)

k = thermal conductivity

ΔT = temperature difference from wall to fluid

$\rho vL/\mu$ = Reynolds number

$c_p\mu/k$ = Prandtl number

For engineering purposes, ϕ is taken as an empirical power function of each dimensionless group, giving equations of the form

$$\frac{hL}{k} = K\left(\frac{\rho vL}{\mu}\right)^x\left(\frac{c_p\mu}{k}\right)^y \tag{12-1c}$$

In the turbulent region, the Reynolds index x varies from 0.5 to 0.8 according to the geometry of the system (see Table 12-1), and the exponent on the Prandtl number varies from 0.3 to 0.4. The constant K depends on the geometry of the system and also on the choice of the characteristic length.

The condition for strict thermal similarity in homologous forced-convection systems is therefore equality of Reynolds numbers (Prandtl numbers are by definition equal). In systems that are geometrically similar but not homologous, i.e., that contain different fluids or the same fluid at widely different temperatures, the values of hL/k at equal Reynolds numbers will be proportional to $(c_p\mu/k)^y$, where y may be taken as 0.4 for fluid heating, 0.3 for fluid cooling. McAdams (161) gives Prandtl numbers for the common gases and for a number of liquids at different temperatures. Since the Prandtl number enters to a low power, it need not be known with great accuracy and, for organic liquids, may be estimated by an empirical relation derived by Denbigh (150)

$$\log\frac{c_p\mu}{k} = \frac{0.2\,\Delta H}{RT} - 1.8 \tag{12-2}$$

where ΔH = molal heat of vaporization at atmosphere boiling point, cal/g mole

T = absolute temperature, °K

R = gas const, cal/(g mole)(°C) = 1.987

For all permanent gases, the Prandtl number may be taken as constant; i.e., in geometrically similar systems at the same Reynolds number all gases will give substantially the same Nusselt number.

The similarity condition of constant Reynolds numbers precludes large-scale ratios owing to the excessive fluid velocity and power consumption required in the model, so that if a forced-convection coefficient is the rate-determining factor, the pilot-scale heat exchanger must be an element or model element of the large-scale unit with a scale ratio not far from unity. For example, where the large-scale unit is to be of the multitubular type, the experimental unit may consist of a few tubes or even a single tube of approximately the same diameter and l/d ratio. The external jacket of the element should be such as to produce a similar flow pattern outside the tubes (see, for example, Fig. 12-1). The frictional and heat-transfer relations for similar systems in terms of scale ratio are given by Eq. (7-11a), (7-11h), and (7-15a) to (7-15f).

Often it is not possible for the small-scale heat-transfer unit to constitute an element of the full-sized one, for example, when the latter consists of a single jacketed tube or a coil immersed in a tank. The small-scale results have then to be extrapolated to higher Reynolds numbers by methods such as those discussed in Chap. 8. For homologous systems, Eq. (12-1c) may be expressed in the simplified form

$$h = J \frac{v^x}{L^{1-x}} \tag{12-3}$$

where h = mean film coefficient of heat transfer

v = mean fluid velocity

L = a characteristic length, such as pipe diameter or spacing between flat plates

J is a constant for the system embodying the shape factor and physical properties of the fluid at the operating temperatures,

$$J = Kk \left(\frac{\rho}{\mu}\right)^x \left(\frac{c_p \mu}{k}\right)^y \tag{12-4}$$

If desired, J may be determined in the experimental heat exchanger over a range of temperature. The Reynolds index x may also be determined experimentally, or an appropriate value may be selected from Table 8-1.

Where the heat-transfer system consists of a coil immersed in a stirred pot, it is convenient to express the external-film coefficient for geometrically similar systems in terms of the stirrer diameter (D) and rpm (N). The fluid velocity v is replaced by the peripheral speed of the stirrer ND, and Eq. (12-3) becomes

$$h = JN^xD^{2x-1} \tag{12-5}$$

The value of J for a given fluid varies somewhat with temperature. Further, Eq. (12-3) is not dimensionless, and the numerical value of J varies with the units employed. Where the model and prototype heat-transfer units are processing the same fluid over the same temperature range, the value of J is the same for both. Equation (12-3) can then be expressed in terms of ratios of corresponding quantities, and J disappears, leading to the general scale-up equation for heat transfer by forced convection

$$\mathbf{h} = \frac{\mathbf{v}^x}{\mathbf{L}^{1-x}} \tag{12-6}$$

This equation is dimensionless, since each term is a ratio of two quantities of the same kind. The value of the exponent depends upon the fluid regime and the geometry of the system. Table 8-1 lists various experimental values taken from the literature.

Once the value of x is known, Eq. (12-6) enables the film coefficient of heat transfer to be scaled up for any geometrically similar apparatus and any fluid velocity, provided only that the fluid regime remains turbulent in both model and prototype. In the case of a stirred vessel, $\mathbf{v}^x$ is replaced by $(\mathbf{ND})^x$, $\mathbf{ND}$ being the ratio of peripheral stirrer speeds.

Equation (12-6) can be still further simplified if the experimental conditions are chosen so that x, $\mathbf{v}$, or $\mathbf{h}$ disappears. There are four such sets of conditions, namely, *dynamic similarity*, *equality of fluid velocities*, *equality of heat-transfer coefficients*, and *equality of heating rates*. The fluid velocity in the model corresponding to a given velocity in the protype is greatest for dynamic similarity and least for equality of heating rates.

Dynamic Similarity. In forced-convection systems, the criterion of dynamic similarity is equality of Reynolds number. If the same fluid is heated or cooled over the same temperature range in both model and prototype, this requires that the fluid velocity shall vary inversely as the scale ratio, that is, $\mathbf{v} = \mathbf{L}^{-1}$. The exponent x disappears, and Eq. (12-6) becomes

$$\mathbf{h} = \frac{1}{\mathbf{L}} \tag{12-6a}$$

In dynamically similar systems, the fluid-flow patterns are similar at all corresponding points in model and prototype, a condition which is theoretically desirable for the accurate scaling up of heat-transfer coefficients. In practice, dynamic similarity is generally attainable only where L is relatively small, otherwise, the fluid velocity in the model becomes impracticably high. In the case of gases, it may reach the point at which compression effects become appreciable, and the Reynolds number is no longer the sole criterion of similarity. Models specially designed for heat-transfer measurements may occasionally be operated under dynamically similar conditions, but this is seldom practicable where the model heat-transfer unit forms part of a pilot plant.

Equality of Fluid Velocities. Where corresponding fluid velocities or stirrer peripheral speeds are equal in model and prototype, v is unity and Eq. (12-6) becomes

$$h = \frac{1}{L^{1-x}} \tag{12-6b}$$

i.e., heat-transfer coefficients tend to be lower on the large scale, but not so much lower as in the case of dynamic similarity. The condition of equal fluid velocities allows of reasonable velocities in the model apparatus. It is especially advantageous in scaling up over-all coefficients with forced convection in one fluid and natural convection or condensation in the other. This is discussed later.

Equality of Heat-transfer Coefficients. Here the model apparatus is operated to give heat-transfer coefficients equal to those in the prototype. h is unity by definition, and Eq. (12-6) becomes

$$v = L^{(1-x)/x} \tag{12-6c}$$

In stirred vessels, $v = ND$, and $D = L$, so that the ratio of stirrer speeds in rpm is given by

$$N = \frac{1}{L^{(2x-1)/x}} \tag{12-6d}$$

This relation was derived by Rushton for the special case of heat transfer in mixers, but it is generally applicable to all types of heat-transfer equipment.

Equality of heat-transfer coefficients is advantageous where there is some localized condition of scaling, fouling, or corrosoin which depends upon the total heat flux across the transfer surface and which it is desired to reproduce in the model. It calls for lower fluid velocities in the model than in the prototype and is even further removed from dynamic similarity than where the fluid velocities are equal.

Equality of Heating Rates. Green (190) has suggested that a model with geometrically similar heat-transfer surfaces could achieve a heat input or output per unit volume equal to that in the prototype if the heat-transfer coefficient in the model were made low enough to compensate for the increased ratio of surface to volume. The required relation is

$$h = L = \frac{v^x}{L^{1-x}}$$

whence

$$v = L^{(2-x)/x} \tag{12-6e}$$

Or, in terms of stirrer rpm,

$$N = L^{(2-2x)/x} \tag{12-6f}$$

This relation calls for very much lower fluid velocities in the model than in the prototype. The ratio of Reynolds numbers would be equal to $vL = L^{2/x}$ and might bring the fluid flow in the model into the transition or streamline region, in which case Eq. (12-6e) would no longer apply. For this reason, it is generally preferable to operate the model at higher fluid velocities and regulate the heating rate by providing a dummy heat-transfer surface as mentioned earlier.

Example. A liquid reaction mixture is to be preheated with steam in a multitubular heater. Small-scale experiments with a single steam-jacketed tube of $\frac{1}{2}$ in. ID by 5 ft long gave over-all heat-transfer coefficients varying from 69 at 1 fps liquid velocity to 170 at 4 fps. What predictions can be made concerning heat-transfer coefficients in a single-pass multitubular heater having tubes of 1 in. ID by 10 ft long?

The single-jacketed tube constitutes a model element of the multitubular unit to a scale ratio of 2 (the section ratio depends upon the number of tubes in the large unit and is immaterial to the problem).

The steam-side coefficient will be so high compared with the liquid side that a rough value may be taken. Assume 1,000. The coefficient corresponding to the thermal resistance of the tube wall is calculated to be 4,000. Using these values, one finds the liquid-side film coefficients to be 76 at 1 fps and 215 at 4 fps. It follows that

$$h \propto v^{0.75}$$

The general scale equation for the system is therefore, from Eq. (12-6),

$$h = \frac{v^{0.75}}{L^{0.25}}$$

Since $L = 2$ and $L^{0.25} = 1.19$, the liquid film coefficient in the multitubular heater for the same mixture over the same temperature range will be given by

$$h = \frac{76}{1.19} v^{0.75} = 64v^{0.75}$$

Natural Convection

Heat transfer by natural convection without change of phase conforms to the generalized equation (7-19),

$$\frac{hL}{k} = \frac{Q}{kL\,\Delta T} = \phi\left(\frac{\beta g\,\Delta T\,L^3\rho^2}{\mu^2}, \frac{c_p\mu}{k}\right) \tag{12-7}$$

As was shown in chap. 7 [Eq. (7-19e)], the conditions for true thermal similarity in homologous systems necessitate an impracticably high temperature difference in the model. It is therefore best for the small apparatus to constitute a full-scale element of the prototype. Where this is impossible (as, for example, where the heat-transfer system consists of a coil immersed in an unstirred tank), methods of extrapolation must be employed. These are based on the Lorenz equation [Equation (8-6)],

$$\frac{hL}{k} = C \left(\frac{\beta g \, \Delta T \, L^3 \rho^2}{\mu^2} \frac{c_p \mu}{k} \right)^z \tag{12-8}$$

For geometrically similar systems in which the same fluid is heated over the same temperature range, the physical properties cancel out as before, leaving the general scale equation for natural convection,

$$\mathbf{h} = \mathbf{\Delta T}^z \mathbf{L}^{3z-1} \tag{12-9}$$

The value of the exponent z depends upon the numerical value of the dimensionless quantity inside the parentheses in Eq. 12-8, the product of the Grashof and Prandtl numbers. Table 8-2 shows the values given by Fishenden and Saunders (151).

As with forced convection, four scale-up relations will be considered: *true thermal similarity* corresponding to equality of Grashof numbers, *equality of temperature differences, equality of heat-transfer coefficients,* and *equality of heating rates.* The temperature difference in the model corresponding to a given temperature difference in the prototype is greatest for thermal similarity and least for equality of heating rates.

Thermal Similarity. For identical fluids and mean temperatures equality of Grashof numbers requires the product $(\Delta T L^3)$ to be constant, and the scale relation is

$$\mathbf{h} = \frac{1}{\mathbf{L}} \tag{12-9a}$$

This is identical with Eq. (12-6a) for dynamic similarity (since in both cases the Nusselt group hL/k is constant).

To achieve such high heat-transfer coefficients in the model, the temperature difference is required to vary inversely as the cube of the linear scale. This would necessitate either impossibly high temperature differences in the model or uneconomically low ones in the prototype. Hence, for direct scaling up, the conditions of true thermal similarity are not practicable.

Equality of Temperature Differences. When ΔT is constant, Eq.

(12-9) simplifies to

$$h = \frac{1}{L^{1-3z}} \tag{12-9b}$$

Equality of temperature differences is the most usual relationship between pilot and full-scale plant. As for forced convection at equal fluid velocities, film coefficients of heat transfer tend to be lower on the large scale than in the model.

Equality of Heat-transfer Coefficients. Where h is unity, Eq. (12-9) becomes

$$\Delta T = L^{(1-3z)/z} \tag{12-9c}$$

Equality of heat-transfer coefficients may occasionally be preferable to equality of temperature differences where there are local effects at the heating or cooling surface which depend upon the total heat flux.

Equality of Heating Rates. To secure equal rates of heat input or removal per unit volume under any of the foregoing conditions while preserving geometrical similarity, it is necessary to provide a proportion of dummy heat-transfer surface as in forced convection. Alternatively, the heat-transfer coefficient could theoretically be reduced in proportion to the linear scale,

$$h = L = \frac{\Delta T^z}{L^{1-3z}}$$

whence $\qquad\qquad \Delta T = L^{(2-3z)/z} \tag{12-9d}$

The practical utility of this relation is doubtful because of the diminution in value of the Lorenz exponent z at low Grashof numbers.

Condensation

Industrial condensers may have either vertical or horizontal tubes. Evaporative condensers generally have vertical tubes with the vapor inside, but otherwise horizontal tubes with the vapor outside are normally preferred because of the higher heat-transfer coefficients obtained. An element of a vertical condenser may consist of a single vertical tube and jacket (like a large-sized Liebig condenser). An element of a horizontal condenser cannot have less than one complete vertical row of tubes, since the heat-transfer coefficient depends partly upon the number of tubes in the row. The profile of the enclosing chamber should be such as to produce a similar flow pattern outside the tubes, e.g., by projections on the walls to represent adjacent tubes as in Fig. 12-1. The horizontal tubes need not be as long as in the prototype, since the condensing film coefficient is not appreciably affected by tube length, but they should have the same outside diameter. Heat-transfer coefficients determined in a full-scale element are directly applicable to the prototype condenser operating under the same conditions.

A dimensionless equation for the film coefficient of heat transfer from a condensing vapor to a cold surface was derived by Nusselt (164) on the assumption that the entire thermal resistance was due to a film of condensate completely covering the heat-transfer surface and descending in streamline motion under the influence of gravity. The equation is

$$\frac{h^4 L \mu \, \Delta T}{k^3 \rho^2 g \lambda} = \text{const} \quad (12\text{-}10a)$$

or $\quad h = K \left(\dfrac{k^3 \rho^2 g \lambda}{L \mu \, \Delta T} \right)^{\!\frac{1}{4}} \quad (12\text{-}10b)$

where h = mean film coefficient of heat transfer

$k,\ \rho,\ \mu$ = thermal conductivity, density, viscosity of condensate film

λ = latent heat of vaporization

ΔT = mean temperature difference from vapor to tube wall

L = (1) length of vertical tubes, (2) nd for horizontal tubes, where d = tube diameter, n = number of tubes in each vertical row

K = a const = (1) 0.943 for vertical tubes, (2) 0.725 for horizontal tubes

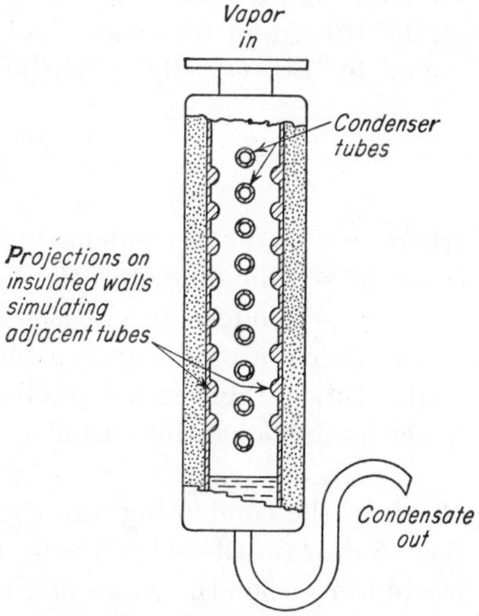

Fig. 12-1. Diagrammatic section of an experimental element of a horizontal condenser.

The dimensionless group on the left-hand side of Eq. (12-10a) is known as the *McAdams*, or *condensation*, group. For homologous systems, the condition for thermal similarity is

$$h^4 L \, \Delta T = h^4 nd \, \Delta T = \text{const}$$

or $\quad \mathbf{h} = \left(\dfrac{1}{\mathbf{L} \, \mathbf{\Delta T}} \right)^{\!\frac{1}{4}} = \left(\dfrac{1}{\mathbf{nd} \, \mathbf{\Delta T}} \right)^{\!\frac{1}{4}} \quad (12\text{-}10c)$

This is a feasible relationship for model experiments involving similarity on the vapor side only. As long as the product $L \, \Delta T$ is constant, the film coefficient h should be constant. For equal temperature differences in model and prototype, the relation between scale ratio and heat-transfer coefficient simplifies to

$$\mathbf{h} = \frac{1}{\mathbf{L}^{0.25}} \qquad (12\text{-}10d)$$

Once again, heat-transfer coefficients on the large scale tend to be lower than in model equipment, provided that the motion of the condensate film is streamline in both cases. In long vertical-tube condensers, the condensate flow may become turbulent toward the bottom of the tube, and film coefficients will be somewhat higher than predicted by Eq. (12-10d). Turbulent flow of condensate sets in when

$$\frac{4w}{\pi\,d\mu} > 2,100$$

where w = lb/hr of condensate
  d = tube diameter, ft,
  μ = condensate viscosity in consistent units

The above group is thus a modified Reynolds number. For horizontal tubes, there is no practical limit to the scale ratio, since d is never large enough for turbulence to set in.

At high condensation rates, the vapor velocity can have a marked effect on the condensing-film coefficient, increasing it when the vapor flow is downward and reducing it (by increasing the thickness of the liquid film) when the vapor flow is upward as in a reflux condenser. In the design of a condenser on paper, this factor may have to be allowed for in calculating the condensing-film coefficient, but in scaling up heat-transfer coefficients from a geometrically similar model in accordance with Eqs. (12-10c) or (12-10d) the effect is largely self-compensating, since for equal values of $L\,\Delta T$ the mass velocity of vapor across a given section of a geometrically similar model is equal to that in the prototype.

For equal film coefficients on the small and large scale, Eq. (12-10c) becomes

$$\mathbf{\Delta T} = \frac{1}{\mathbf{L}} \qquad (12\text{-}10e)$$

From the two forms of Nusselt's equation (12-10a), it is possible to derive a relation enabling the film coefficient in a horizontal multitubular condenser to be predicted from experiments with a single short vertical tube (short enough to ensure streamline flow). The relation is

$$\frac{h_h}{h_v} = 0.769 \left(\frac{L}{nd}\right)^{\frac{1}{4}} \qquad (12\text{-}10f)$$

where h_h, h_v = film coefficients for horizontal, vertical tubes

$\qquad L$ = length of vertical tube, ft

$\qquad d$ = diameter of horizontal tube, ft

$\qquad n$ = number of horizontal tubes in a vertical row

The foregoing equations all refer to normal film condensation. When steam condenses on a surface smeared with an oily or nonwetting agent, dropwise condensation takes place with film coefficients up to eight or ten times those for film condensation. Usually the effect is only temporary until the contaminant is washed off, though attempts are being made to sustain dropwise condensation by adding volatile agents to the steam. In general, dropwise condensation is to be regarded as a passing effect, and condensers are designed on the basis of normal film condensation. In scaling up the performance of a model condenser, one should be sure that the surface was not contaminated with any substance that might have caused dropwise condensation while the measurements were being made; otherwise, the scaled up coefficients could be a great deal too high.

Ebullition

Heat transfer with ebullition may take place inside or outside tubes and coils or on plane or curved surfaces such as the walls of jacketed vessels. For a given liquid, the film coefficient h depends upon the temperature difference, pressure, roughness of the heating surface, interfacial tension between surface and liquid, and depth of submergence. The overriding factor is the temperature difference. As it increases, the total heat flux = $h \Delta T$ rises steeply to a maximum value in the region of $\Delta T = 40 - 80°F$, after which a further increase in ΔT causes it to diminish owing to blanketing of the heat-transfer surface by a film of vapor.

There is at present no satisfactory theory or mathematical analysis of boiling. For very low temperature differences, the heat-transfer rate has been found to conform to the equation for natural convection [Eq. (12-8)], and the scale effect is therefore represented by Eq. (12-9) (151). At ordinary boiling rates, neither the size, shape, or orientation of heating surfaces has much effect upon the rate of heat transfer. Beyond a certain depth, from 5 to 20 in. in different evaporators, the apparent heat-transfer coefficient appears to diminish somewhat with increasing submergence (161). The empirical dimensionless correlation of Insinger and Bliss (152) contains neither shape factor nor linear dimensional term.

It would appear that film coefficients of heat transfer to boiling liquids may be expected to be of the same order of magnitude in small-

and large-scale apparatus irrespective of geometry or scale ratio. There is, however, a slight tendency for coefficients to diminish with increasing size of plant at very low heat-transfer rates and beyond certain submergence depths.

Over-all Coefficients

Pilot-scale heat-transfer units are rarely equipped with the delicate surface thermocouples required for the direct measurement of film coefficients. The experimental heat-transfer data obtained from pilot-plant operation will therefore normally consist of over-all coefficients. In certain cases, where there is forced convection on one side and the value of the Reynolds-number exponent x is known, the over-all coefficient may be resolved into film coefficients by means of a Wilson plot (174). It is, however, a great practical advantage if the experimental conditions are so chosen that the over-all heat-transfer coefficient can be directly scaled up as such.

In what follows, the thermal resistance of the solid partition will be neglected. Where it is not negligible, its thickness must be adjusted so that the ratio of wall resistance in prototype and model is the same as the ratio of film resistances. This will generally call for somewhat thinner walls in the model than in the prototype, but not as thin as would correspond to strict geometrical similarity.

For the direct scaling up of combinations of film coefficients for forced convection, natural convection, condensation, and boiling, the requirement is that the ratio h of film coefficients in the prototype and model shall be the same for both films. This can be achieved in a number of ways in accordance with the foregoing equations, but the following conditions are recommended as the simplest and most convenient:

1. Small- and large-scale heat-transfer units geometrically similar or approximately so, with dummy surface on the small scale as required to regulate the heating rate

2. Fluid velocities or peripheral stirrer speeds equal on the small and large scale

3. Temperature differences equal on the small and large scale

Under these conditions the scale-up relation for film coefficients in forced convection is given by Eq. (12-6), in natural convection by Eq. (12-9), and in condensation by Eq. (11-10c). It so happens that in all these equations the exponent on L is of the same order of magnitude. No equation was given for boiling, but the indications were that the trend is in the same direction, i.e., lower film coefficients in larger plant. This is shown in Table 12-1.

The scale-up relation for over-all coefficients under the recommended conditions is

$$U = \frac{1}{L^s} \tag{12-11}$$

where s is chosen according to the predominant heat-transfer mechanism.

Edgeworth Johnstone (156) made a direct comparison of over-all heat-transfer coefficients in a model steam-jacketed pan and a proto-

TABLE 12-1. EXPONENTS ON L UNDER RECOMMENDED SCALE-UP CONDITIONS

Mechanism	Exponent on $L = s$	Numerical range
Forced convection.............	$1 - x$	0.2–0.5
Natural convection...........	$1 - 3z$	0.01–0.55
Condensation................	0.25	0.25
Ebullition...................	Probably positive	

TABLE 12-2. RECOMMENDED MEAN EXPONENTS ON L [EQUATION 12-11]

Controlling mechanism	s
Turbulent flow inside tubes Vigorous natural conveetion Condensation Boiling	0.25
Flow across tube banks Vigorously stirred vessels with coils or jackets Moderate natural convection	0.33
Flow across finned tubes or through broken solids Jacketed vessels with slow stirring Mild natural convection	0.5

type to a linear scale ratio of 6.5 (volume ratio 275:1). Under the recommended conditions, the experimental value of the exponent on **L** was $s = 0.23$ for natural convection and $s = 0.46$ with slow stirring (a combination of natural and forced convection), thus verifying the relations derived from heat-transfer equations.

In the absence of other data, it is suggested that the values of Table 12-2 be used.

Where both film coefficients are likely to be of the same order of magnitude but have different exponents on **L**, an intermediate value of s may be taken. Strictly speaking, Eq. (12-11) applies only to geometrically similar systems, but even where they are not geometrically similar, provided that the model and prototype units are of similar design and that they are compared under the conditions recommended, a direct scale-up by means of Eq. (12-11) is likely to be more accurate than a calculation from generalized correlations. The greater the departure from geometrical similarity, however, the greater the possibility of error.

Since the value of s hardly ever exceeds 0.5, it is a reasonably safe design rule to make fluid velocities (or peripheral stirrer speeds) and temperature differences equal in model and prototype and then take over-all heat-transfer coefficients as inversely proportional to the square root of the linear scale.

Example. A chemical reaction takes place in an experimental reaction vessel 18 in. in diameter and having a stirrer. The stirrer speed is 800 rpm, and it is found that the reaction temperature can be held at 85°C by means of a cooling coil containing 2 ft of $\frac{1}{2}$-in. OD tubing through which water passes at a mean velocity of 3 fps. The inlet and outlet temperatures of the cooling water are 20 and 25°C, respectively. How much heat-transfer surface should be provided in a geometrically similar vessel 6 ft in diameter to control the same reaction at the same temperature?

The linear scale ratio $L = 6/1.5 = 4.0$.

Assuming geometrically similar stirrers, the speed of that in the larger vessel should be $800/4 = 200$ rpm.

The large-scale cooling coil will be of $\frac{1}{2} \times 4 = 2$-in. OD tubing in which the water velocity will be 3 fps as in the model. The ratio of heat-transfer coefficients is obtained from Eq. (12-11), taking an intermediate value of s, say, 0.3.

$$U = \frac{1}{4}^{0.3} = 0.66$$

At the same water velocity, the large coil uses only 16 times the cooling water to remove 64 times the heat; so the temperature rise will be 4 times that in the model, from 20 to 40°C. The log mean temperature difference is thus reduced from 62.5 to 54.5, and the ratio $\Delta T = 0.87$.

The cooling surface found adequate in the experimental vessel was 0.52 ft². That required in the large vessel with a stirrer speed of 200 rpm and a cooling-water velocity of 3 fps in the coil is

$$0.52 \, \frac{L^3}{U \, \Delta T} = 0.52 \times 4^3/0.66 \times 0.87 = 57.9 \text{ ft}^2$$

corresponding to 110 ft of 2-in. OD tubing.

The above calculation neglects the heat lost by natural convection from the wall of the vessel to the air. For this assumption to be justified, the exterior of the experimental pan should be insulated to reduce the surface heat loss per square foot to not more than $1/L$ times that from the bare wall of the large pan.

An alternative method of scaling up over-all heat-transfer coefficients without measuring wall temperatures is to separate the film coefficients by means of a Wilson plot (174). First, values of the Reynolds index for both sides of the heat exchanger must be assumed in the light of Table 12-1 or other information. Call these x and x'

for the inside and outside streams, respectively. Then:

$$h_i = J_i \frac{v_i^x}{L_i^{1-x}} \qquad h_o = J_o \frac{v_o^{x'}}{L_o^{1-x'}}$$

$$\frac{1}{U} = \frac{1}{h_i} + \frac{1}{h_w} + \frac{1}{h_o}$$

$$= \frac{L_i^{1-x}}{J_i} \frac{1}{v_i^x} + \frac{1}{h_w} + \frac{L_o^{1-x'}}{J_o} \frac{1}{v_o^{x'}} \qquad (12\text{-}12)$$

where U = over-all heat-transfer coefficient

L_i, L_o = characteristic lengths or diameters for inside and outside fluid stream

v_i, v_o = inside and outside fluid velocity

J_i, J_o = inside and outside values of the constant J

h_w = conductance per unit area of the wall

Let v_o be kept constant and U measured for several values of v_i, all in the range of turbulent flow. If $1/U$ is then plotted against $1/v_i^x$, a straight line should be obtained, the slope of which corresponds to L_i^{1-x}/J_i and the intercept at $v_i = \infty$ to $L_o^{1-x'}/J_o v_o^{x'} + 1/h_w$.

The value of h_w can be calculated from the thickness and thermal conductivity of the wall. Thus J_L and J_o can be determined and empirical equations similar to Eq. (12-3) or (12-5) set up for both sides of the exchanger. The assumption that h_o remains constant while h_i is varied is not strictly correct, since the film temperature and viscosity in the outside stream will alter slightly with changes in h. The J values determined by means of the Wilson plot should therefore be checked against the direct scale-up at equal velocities according to Eq. (12-6) and if necessary adjusted to conform with the latter.

As an alternative to the Wilson plot, the film coefficients may be separated algebraically by the method of regression analysis. In a heat exchanger where the velocity on one side, say, v_i, is varied, while that of the other, say, v_o, is held constant, we can write

$$Z = A + Bp \qquad (12\text{-}13)$$

where
$$Z = \frac{1}{U}$$

$$A = \frac{1}{h_w} + \frac{1}{h_o} \text{ (const)}$$

$$= \frac{1}{h_w} + \frac{L_o^{1-x'}}{J_o v_o^{x'}}$$

$$p = \frac{1}{v_i^x}$$

$$B = \frac{L_i^{1-x}}{J_i}$$

The problem is, from a series of experimental values of Z versus p and knowing or assuming the value of x, to find A and B and thus obtain the values of Z and p. Let $(Z_1 p_1)$, $(Z_2 p_2)$, $(Z_3 p_3)$ represent corresponding sets of observed values of Z and p.

$\bar{Z}$ = arithmetical mean of Z_1, Z_2, Z_3

$\bar{p}$ = arithmetical mean of p_1, p_2, p_3

The regression formula is

$$Z = \bar{Z} + \frac{\Sigma(p-\bar{p})(Z-\bar{Z})}{\Sigma(p-\bar{p})^2} (p-\bar{p}) \tag{12-14}$$

whence A and B can be determined. Knowing the conductivity of the wall, one can calculate the value of h_w and obtain J_o and J_i.

Example. In a double-pipe heat exchanger, the inside diameter of the inlet pipe is $1\frac{7}{8}$ in.; the outside diameter of the inner pipe and the inside diameter of the outer pipe are 2 and 3 in., respectively. With the outer velocity held constant at 9 fps and the inner velocity varied, the following set of over-all heat transfer coefficients was measured:

v_i	U
10	305
6	271
3	217

For this system, the value of the Reynolds index x can be taken as 0.8 on both sides. As characteristic lengths, we choose the inside diameter of the inner pipe, $L_i = 0.146$ ft, and the hydraulic mean radius of the annular space, $L_o = 0.0208$ ft. Tabulating corresponding values of Z and p,

U	$\frac{1,000}{U} = Z$	v_i, fps	$v_i^{0.8}$	$\frac{1,000}{v_i^{0.8}} = p$
305	3.28	10	6.31	159
271	3.69	6	4.19	238
217	4.60	3	2.41	415

Values of $1/U$ and $1/v_i$ have been multiplied by 1,000 for convenience.

$$\bar{Z} = 3.86 \qquad \bar{p} = 270.8$$
$$\Sigma(p-\bar{p})(Z-\bar{Z}) = 177.5 \qquad \Sigma(p-\bar{p})^2 = 34,454$$

whence
$$Z = 2.46 + 0.00514p$$
$$\frac{1}{U} = 0.00246 + \frac{0.00514}{v_i^{0.8}}$$

Substituting $L_i^{0.2} = 0.6789$,

$$\frac{1}{U} = 0.00246 + \frac{1}{132} \frac{L_i^{0.2}}{v_i^{0.8}}$$

whence $J_i = 132$.

$$\frac{1}{h_w} + \frac{0.460}{J_o v_o^{0.8}} = 0.00246$$

and

$$\frac{1}{v_o^{0.8}} = \frac{1}{5^{0.8}} = 0.276$$

h_w is found by calculation to be 1,270, whence $J_o = 195$.

For simplicity, the foregoing example is based upon only three pairs of observed values, but this would not normally be considered sufficient for a regression analysis. The number of pairs of values should preferably be 10 or more.

Where it is not feasible to hold one fluid velocity constant while varying the other, the regression equation becomes

$$Z = A + Bp + Cq \tag{12-15}$$

where $q = 1/v_o^{x'}$.

This can be solved by the method of multiple regression. The constants B and C are found by solving the simultaneous equations

$$\Sigma(Z\text{-}\bar{Z})(p\text{-}\bar{p}) = B\Sigma(p\text{-}\bar{p})^2 + C\Sigma(p\text{-}\bar{p})(q\text{-}\bar{q}) \tag{12-16a}$$
$$\Sigma(Z\text{-}\bar{Z})(q\text{-}\bar{q}) = B\Sigma(p\text{-}\bar{p})(q\text{-}\bar{q}) + C\Sigma(q\text{-}\bar{q})^2 \tag{12-16b}$$

Substituting the values of B and C in the least-squares equation,

$$Z\text{-}\bar{Z} = B(p\text{-}\bar{p}) + C(q\text{-}\bar{q}) \tag{12-16c}$$

and solving for Z gives the value of A. For the proof of the method see statistical works, for example, Davies (132).

There is a potential source of error in all of the foregoing extrapolation methods in that for any fluid velocity below that corresponding to equal Reynolds numbers the change in fluid temperature is greater in the model than in the prototype owing to the higher ratio of surface to volume. There may thus be a difference between the mean temperature at which the heat-transfer coefficient is determined and that to which it is extrapolated. Generally, the effect is negligible, but if necessary it can be counteracted by modifying the inlet temperatures so that the average temperature of each fluid is the same in model and prototype. Another expedient is to have a proportion of dummy heat-transfer surface in the model in order to maintain the same ratio of active surface to volume as in the prototype while preserving geometrical similarity. This is sometimes desirable in chemical reactors, where the rate of heat transfer per unit volume is important (see Chap. 15).

Experimental or pilot-plant heat exchangers from which it is proposed to scale up heat-transfer rates should be provided with inlet and outlet thermometers on both sides, also flowmeters on both fluid streams. It should be possible to vary both rates of flow or hold either one constant at a given rate. Thermocouples in the metal wall are useful where they can conveniently be fitted but are not necessary when the methods discussed above can be applied.

All of the foregoing scale-up methods assume that the fluid regime is turbulent in the model as well as in the prototype, and this should in all cases be verified by calculating or roughly estimating the Reynolds number.

SYMBOLS IN CHAPTER 12

$A = 1/h_w + 1/h_o$

$B = L_i^{1-x}/J_i$

C = const in Eq. (12-8)

c_p = specific heat at constant pressure

D = diameter of stirrer

d = diameter of pipe

g = acceleration of gravity

h = film coefficient of heat transfer

ΔH = molar heat of vaporization at atmospheric boiling point, cal/g mole

J = heat-transfer parameter defined by Eq. (12-4)

K = const

k = thermal conductivity (of fluid)

L = linear dimension

l = length of pipe

N = rotational speed of stirrer, rpm

n = number of horizontal tubes in a vertical row

$p = 1/v_i^x$

Q = total heat transferred in unit time

$q = 1/v_o^{x'}$

R = gas const, cal/(g mol)(°C)

s = empirical exponent on L in Eq. (12-11)

T = absolute temperature

ΔT = temperature difference

U = over-all heat-transfer coefficient

v = fluid velocity

w = lb/hr of vapor condensing on a vertical tube

x = exponent on Reynolds number (Reynolds index)

y = exponent on Prandtl number

$Z = 1/U$

z = exponent in Eq. (12-8)

β = coefficient of cubical expansion of fluid

λ = latent heat of vaporization

μ = fluid viscosity

ρ = fluid density

Subscripts

i, o = inside, outside tubes

w = relating to wall of tube

h, v = horizontal, vertical tubes

CHAPTER 13

PACKED TOWERS

Note. In this chapter, following current usage, the symbol L denotes not linear dimension but superficial liquid mass velocity in pounds per square foot per hour. Similarly, L would denote the ratio of liquid mass velocity in the prototype to that in the model. Linear dimensions are denoted by d, packing-element diameter in feet; a, specific surface of packing in square feet per cubic foot; D, tower diameter in feet; and Z, packed height of tower in feet.

Towers or columns are among the most commonly used devices for contacting two immiscible fluids. Their principal fields of application are in distillation, gas absorption, evaporative cooling, and liquid extraction. The fluid flow in towers is generally countercurrent, though not necessarily so except in the case of fractional distillation. The principal types of tower are *wetted wall, spray, packed, baffle plate, sieve plate, perforated plate*, and *bubble plate.*

Wetted-wall towers have hitherto been used only for research, for which they are particularly suited because their cross-section and interfacial area are known and constant. Industrial wetted-wall towers containing a bundle of vertical tubes have been tried experimentally. They have the advantage of a low pressure drop per theoretical stage, but their capacity is also low, and they have not come into general use. The various types of plate column present no particular problems in scaling up. Large plate columns correctly loaded tend to have higher plate efficiencies than smaller ones of similar design, so that it is generally safe to scale up a pilot-plant column plate for plate. In any case, the plate efficiency of an experimental column is easily determined and compared with typical full-scale values. About spray towers there is not enough information to suggest formal rules for scaling up. Spray towers are not used where mass-transfer performance is critical.

The scaling up of packed towers is a real problem because the performance of a given tower can vary greatly with the conditions of operation and the generalized design correlations are more than usually inexact. A great deal of work has been published on packing performance, and in spite of conflicting similarity criteria it is possible to formulate empirical rules for scaling up most types of operation which,

151

for a particular case, are likely to be more reliable than generalized correlations.

In a countercurrent packed tower, the heavier fluid is introduced at the top and descends by gravity against a rising stream of the lighter fluid. Either phase can be dispersed, but in gas-liquid towers it is always the liquid that is dispersed because of the excessive power consumption when gas is bubbled through a tall column of liquid. In liquid-liquid towers, the lighter fluid is often dispersed and caused to rise in droplets through the packing. The relation of mass-transfer rates to the geometry and dimensions of the packing depends very much on whether or not the packing is preferentially wetted by the disperse phase. If it is so wetted, the interfacial area may be expected to bear some relation to the specific surface of the packing. If the disperse phase does not wet the packing but moves through it in bubbles or droplets, the size and shape of the packing elements are likely to have much less effect. In gas-liquid systems such as absorption and distillation columns, the descending liquid always wets the packing, and it has been found that where it is prevented from doing so the efficiency of contact is reduced (126). Liquid-extraction columns, on the other hand, appear to be more efficient when the disperse phase does not wet the packing (180, 183, 184).

Where the packing is not wetted, its shape and size have relatively little influence on the performance of the tower. Treybal states that in liquid extraction the HTU (height of a transfer unit) for packed towers is usually only 20 to 25 per cent lower than for empty spray towers (185), which could be accounted for by the more tortuous path which the droplets follow in a packed tower. The principal geometrical factors appear to be the tower height and the diameter or specific surface of the dispersed droplets or bubbles. In liquid extraction, the droplet size tends to be constant for a given system above a critical minimum packing diameter (182). In unpacked spray towers, on the other hand, the droplet size is a function of spray-nozzle dimensions. Ideally, it might be possible in a small-scale spray tower to reduce the droplet diameter and so achieve the same over-all composition change in a smaller height than on the full scale, but the experimental data at present available do not afford a basis for scale relations on these lines.

Towers in which the disperse phase does not wet the packing have one experimental advantage: provided that the wall of the tower is not wetted either, such towers suffer less from wall effects than wetted packing towers, and it appears from the literature that plant-scale mass-transfer coefficients can sometimes be satisfactorily predicted from experimental towers 2 or 3 in. in diameter. In liquid extraction, according to Lewis, Jones, and Pratt (182), the packing diameter in the experimental tower should be not less than $\frac{1}{2}$ to $\frac{3}{4}$ in. Above this diameter, the droplet size is constant; below it, contrary to expectation, the droplet size increases.

In what follows, it will be assumed that the main object is to reproduce in a large-scale tower the performance of a given pilot-scale tower, employing packing elements of the same shape, and fluids of the same initial and final compositions in the same proportions and at the same temperatures and pressures. That is, the discussion will be limited to what have been called *homologous* systems. The physical properties of the fluids under operating conditions are in general assumed to be unknown.

The performance of a packed tower is generally evaluated in terms of either the height of packing required to effect a given change (HETP, height equivalent to a theoretical plate, or HTU, height of a transfer unit*) or, in the case of absorption and extraction, the over-all mass-transfer coefficient on a volume basis ($K_g a$ or $K_l a$). Film coefficients of mass transfer, though theoretically more correct, are difficult to use owing to uncertainty as to the degree of wetting and the interfacial area between the phases. For scale-up purposes, the geometrical conception of HETP or HTU is the more convenient. Although in general the HETP and HTU in a given system are not equal, it may be assumed that the ratio of the HETP in prototype and model is equal to the ratio of HTU. This ratio of corresponding heights will be denoted by **Z**. In accordance with the convention adopted throughout this book, the value for the prototype is understood always to form the numerator of the ratio.

For homologous systems, ratios of HETP or HTU may be converted into ratios of volume-basis mass-transfer coefficients as follows:

$$\mathbf{K}_g \mathbf{a} = \frac{\mathbf{G}}{\mathbf{Z}}$$

$$\mathbf{K}_l \mathbf{a} = \frac{\mathbf{L}}{\mathbf{Z}}$$

where $\mathbf{K}_g\mathbf{a}$, $\mathbf{K}_l\mathbf{a}$ = ratios of volume-basis coefficients on a gas or liquid

* The height equivalent to a theoretical plate is defined as the height of a section of packing such that the vapor leaving the top of the section has the same composition as the vapor in equilibrium with the liquid leaving the bottom of the section. The height of a transfer unit is defined as the height of a section of packing producing a change in vapor composition equal to the mean difference between the actual and equilibrium vapor compositions over the section.

basis, respectively, and **G, L** = ratios of gas and liquid flows per unit cross section.

For liquid extraction, (G,K_ga) and $L,K_la)$ may be taken to refer to the continuous and disperse phases, respectively.

Similarity Criteria

Packed towers constitute one of the most intractable cases of a mixed regime. In a gas-liquid tower, the flow pattern of the gas phase is controlled by the viscosity of the gas and the pressure difference; the flow of liquid is controlled by liquid viscosity and the force of gravity. In a liquid–liquid-extraction tower, the flow of liquid phase is controlled by viscosity and pressure difference, that of the disperse phase by gravity and interfacial tension. In both types of tower, the similarity criteria for the two phases are incompatible for homologous systems when the linear scale is changed.

The criterion for liquid flow in a packed tower may be derived from Nusselt's equation for the isothermal streamline flow of liquid down a vertical wall (164),

$$\frac{w}{l} = \frac{\rho_l^2 g \delta^3}{3\mu_l} \tag{13-1a}$$

where w = liquid rate, lb/hr

l = length of wetted perimeter, ft

δ = thickness of streamline film, ft

ρ_l, μ_l = density, viscosity of liquid at the film temperature in pound-foot-hour units

g = acceleration of gravity, ft/hr^2

For a completely wetted tower packing, this equation may be written

$$Ld = \frac{1}{C} \frac{\rho_l^2 g \delta^3}{\mu_l} \tag{13-1b}$$

where L = superficial mass velocity of the liquid, lb/ft^2-hr

d = diameter or equivalent diameter of packing elements, ft

C = a const

which gives the dimensionless equation

$$\frac{\delta^3}{d^3} = C \frac{L\mu_l}{\rho_l^2 d^2 g}$$

For kinematic similarity, δ/D is constant, whence

$$\frac{L\mu_l}{\rho_l^2 d^2 g} = \text{const} \tag{13-1c}$$

This is the similarity criterion for the liquid phase. For the gas phase, the similarity requirement is a constant Reynolds number,

$$\frac{Gd}{\mu_g} = \text{const} \tag{13-2}$$

where G = superficial mass velocity of the gas, lb/ft^2-hr, and μ_g = viscosity of gas.

For chemical similarity, there is a further requirement of constant liquid/gas ratio,

$$\frac{L}{G} = \text{const} \tag{13-3}$$

In homologous systems, these three requirements reduce to

$$\frac{L}{d^2} = \text{const}$$
$$Gd = \text{const} \tag{13-4}$$
$$\frac{L}{G} = \text{const}$$

which are incompatible when d is varied.

Analysis of the similarity requirements for liquid-liquid towers shows that here also the criteria are incompatible for homologous systems. On theoretical grounds, it might appear impossible to predict the performance of large-scale packed towers by means of experiments with models or to apply the results obtained with any one size of packing to another size. Yet in practice small-scale experiments are frequently used to furnish plant-design data, and results with packing elements of different shapes and sizes have been successfully correlated on a common basis. *Every generalized correlation embodies an implicit rule for scaling up results.* When one is concerned with processing the same chemical substances and maintaining the same physical conditions on the small and large scales, as is generally the case in pilot-plant experimentation, there is no need to insert all the physical properties in the generalized correlation and work out the result. It is simpler to extract from the correlation the empirical scale-up rule and use that directly.

The rest of this chapter will be devoted to a survey of published experimental results with packed towers and an attempt to extract from the many generalized correlations a few reliable scale-up rules.

Liquid Distribution

It is well known that in wetted packing towers there is a tendency for the descending liquid to drift toward the walls. Kirschbaum (218) and Weimann (227) investigated the flow pattern in ring-packed towers and found that this tendency increases with increasing ring diameter for a given tower diameter and decreases with increasing tower diameter for a given ring diameter. This suggests that for similar liquid-flow pattern there should be some relation between packing diameter and tower diameter, and Weimann recommends that the tower diameter should be not less than 25 times the ring diameter. The same workers found that the liquid rate has little effect on distribution and that liquid spreads toward the walls for a certain distance from the tower top and thereafter the distribution remains constant. The latter observation was confirmed by Scott (224).

Baker, Chilton, and Vernon (212) found that vapor velocity does not sensibly affect the liquid distribution until near the loading point, when the distribution becomes more uniform. They recommend that the tower diameter should be not less than 15 times the packing diameter for rings or 8 times the packing diameter for other shapes. The distribution graphs plotted by these workers suggest that for random packings with a central liquid inlet at the tower top the distribution becomes constant after a distance of about 8 tower diameters from the top irrespective of the packing diameter. Stacked packing rings, on the other hand, are not self-distributing, and unless the liquid is uniformly distributed at the top, it is not likely to become so lower down (122).

Under operating conditions, the packing in a tower is rarely wetted completely, and the percentage wetted decreases with the liquor rate and packing diameter. Mayo, Hunter, and Nash (221) found that, for geometrically similar random ring-packed towers, equal percentages of the packing surface were wetted when the liquid mass velocities were approximately in proportion to the linear scale. The percentage of surface wetted was independent of gas velocity. The liquid distribution across the tower was found to be Gaussian, an observation that was confirmed by Tour and Lerman (226) for low liquid rates. Near the flooding rate, the distribution approaches uniformity.

Flooding Point

For every liquid rate, there is a gas rate above which the pressure drop rises steeply, and a layer of liquid appears on top of the packing. The greater the liquid rate, the lower is the gas rate at which this occurs. Flooding represents the limiting gas velocity at which it is possible to maintain countercurrent flow; above the flooding velocity, most of the liquid is blown back out of the tower top. Cocurrent towers do not flood.

Sherwood, Shipley, and Holloway (225) obtained a correlation of flooding rates in gas-liquid towers which for homologous systems at constant gas/liquid ratio reduces to

$$\frac{v_g{}^2}{d} = \text{const} \tag{13-5}$$

or
$$\mathbf{v}_g = \mathbf{d}^{\frac{1}{2}} \tag{13-5a}$$

In correlating packings of different shapes, d is taken as the hydraulic diameter $= 4\epsilon/a$, and v_g is equal to $G/\epsilon\rho g$.

Scale equations identical with (13-5a) result from the correlations of Elgin and Weiss (217), Bertetti (213), Lobo, Friend, Hashmall, and Zenz (220), and Dell and Pratt (216) for gas-liquid columns, and also from those of Dell and Pratt (178) and Crawford and Wilke (177) for liquid-liquid columns. Hence, it appears that for geometrically similar packings of commercial size and in otherwise homologous systems the flooding velocity varies as the square root of the packing diameter. In small laboratory packings, surface-tension effects may cause flooding at velocities lower than those given by this rule. Dell and Pratt (178), extracting various solvents with water, noted a lowering of the flooding velocity below the predicted value with packing elements of $\frac{1}{2}$ in. diameter and less.

PRESSURE DROP

Burke and Plummer (214) found that, for dry packings,

$$\Delta p = B \frac{\rho_g v_g{}^2 a}{\epsilon^3} \tag{13-6}$$

where Δp = pressure drop per foot height of packing
  ρ_g = gas density
  v_g = gas velocity
  a = specific surface of packing
  ϵ = voidage fraction of packing
  B = a friction factor which is an inverse function of the Reynolds number

Chilton and Colburn (215) correlated fluid friction through packed solids by relations which for homologous systems reduce to

$$\Delta p \propto \frac{G^{1.85}}{d^{1.15}} \qquad \text{in the turbulent range} \tag{13-7a}$$

$$\propto \frac{G}{d^2} \qquad \text{in the streamline range} \tag{13-7b}$$

where G = superficial mass velocity of gas and d = diameter of solid particles. Piret, Mann, and Wall (222) found that Chilton and Colburn's correlation applied quite well to ring packings. Leva (219) derived an empirical formula in which

$$\Delta p \propto \frac{G^{1.9}(1 - \epsilon)}{d^{1.1}\epsilon^3} \tag{13-8}$$

where d = diameter of equivalent volume sphere. Sherwood and Pigford (127) recommend a relation which in the turbulent region reduces to

$$\Delta p \propto \frac{G^{1.85}a^{1.15}}{\epsilon^3} \qquad \text{for random packings} \tag{13-9a}$$

$$\propto \frac{G^{1.8}a^{1.2}}{\epsilon^3} \qquad \text{for grids} \tag{13-9b}$$

This gives the pressure drop in the dry packing, which is multiplied by a factor depending on the liquid rate and packing diameter. The gas flow is turbulent when (122):

$$\frac{4G\epsilon}{\mu a} > 100\text{--}300 \qquad \text{for random rings and grids}$$

$$> 500\text{--}1,500 \qquad \text{for stacked rings}$$

As the gas velocity in a packed tower is increased with a constant liquid load, a point is reached, below the flooding point, at which liquid begins to build up in the tower. This is shown by a change in the relation between pressure drop and velocity and is called the *loading point*. Sherwood and Pigford (127) define the loading point as the point at which the pressure drop begins to be proportional to the gas velocity to a power greater than 1.8. Morris and Jackson (122) define the loading point as the point at which the pressure drop varies as the square of the gas velocity, or

$$\frac{dp}{dv_g} \propto v_g$$

In random ring packings, the gas velocity at the loading point is 55 to 75 per cent of the flooding velocity according to the L/G ratio. With Berl saddles, the variation is 67 to 95 per cent. The higher the L/G ratio, the more nearly the loading point approaches the flooding point. Figure 13-1 shows the approximate relation between loading velocity, flooding velocity, and L/G ratio, cross-plotted from curves given in Sherwood and Pigford (127, Figs. 95 to 98), which were derived from published and unpublished data by various workers. For regular packings, e.g., grids and stacked rings, the loading velocity is a lower proportion of the flooding velocity, of the order of 30 per cent. At a given liquid rate, the loading-gas rate in regular packings varies much less with size and type of packing than does the flooding-gas rate.

Kirschbaum has observed in packed distillation columns that for a wide variety of packings and distillates the pressure drop across the

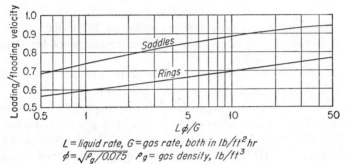

L = liquid rate, G = gas rate, both in lb/ft^2 hr
$\phi = \sqrt{P_g/0.075}$ P_g = gas density, lb/ft^3

Fig. 13-1. Ratio of loading to flooding velocity in packed absorption towers. (*T. K. Sherwood and R. L. Pigford, "Absorption and Extraction," 2d ed., figs. 95–98, McGraw-Hill Book Company, Inc., New York, 1952.*)

packing at the loading point lies between 1.1 and 1.6 mm Hg per foot height of packing (88).

HEIGHT OF PACKING

The height of packing required to effect a given separation is impossible to calculate from theory alone and can be estimated only approximately from empirical design data. It varies with the gas and liquid velocities, the flow pattern in the tower, and whether the gas film or the liquid film controls the rate of mass transfer. The evidence on the effect of packing diameter on HTU or HETP is sometimes conflicting. In reviewing the literature, it will be convenient to consider distillation, gas absorption, evaporation, and liquid extraction separately.

Distillation

Peters (89) and Kirschbaum (88) state that the HETP of random ring packing varies directly as the diameter of the rings = d. Data tabulated by Jacobs (87) suggest that the HETP varies as $d^{0.85}$. On the other hand, Furnas and Taylor (84) found that, between ⅜ and 2 in., packing diameter had relatively little effect on HETP. Robinson and Gilliland (90) give a tentative formula for large towers in which packing diameter does not appear, but the HETP varies directly as the tower diameter and as a compound function of the gas and liquid rates. For homologous systems, it may be written

$$\text{HETP} = D \left(aG^{0.2} + b \frac{G}{L^{0.75}} \right) \tag{13-10}$$

where D = tower diameter, ft
$\quad G$ = superficial mass velocity of vapor, lb/ft²-hr
$\quad L$ = superficial mass velocity of liquid, lb/ft²-hr
$\quad a, b$ = const embodying the physical properties of the vapor and liquid reflux

According to the empirical correlation of Hands and Whitt (86), the HETP of packed distillation towers varies as $(d/L)^{1/2}$.

This conflict of opinion is probably due to the fact that most workers on distillation in packed towers have not been primarily interested in the effect of varying the packing diameter. Other factors have a much greater effect on HETP, notably the slope of the vapor-liquid equilibrium curve and the ratio of operating to flooding velocity. Both Furnas and Taylor (84) and Duncan, Koffolt, and Withrow (83) found that the over-all HTU in a packed distillation tower increases with the average slope of the equilibrium curve and decreases as the operating rate approaches the flooding point. Kirschbaum found that the HETP had a minimum value at the flooding point, reached a maximum of about 1.3 times this value at 60 per cent of the flooding velocity, then fell to the minimum value again at 10 per cent of the flooding velocity. This has not, however, been confirmed by other workers (83, 84).

Evidently the effects of packing diameter and tower geometry on HETP or HTU may easily be masked by other factors unless homologous systems are compared at equal fractions of their respective flooding velocities. All that the literature reveals is that when a model tower is scaled up the HETP is not likely to increase in a greater proportion than the packing diameter and may vary as a fractional power thereof.

Gas Absorption

Scale-up relations for absorption towers depend on whether the liquid-film resistance or the gas-film resistance is the controlling factor. Sometimes this is evident from the nature of the process; for example, the adiabatic evaporation of water into an air stream is bound to be controlled by the gas-film resistance. In other cases, the control must be determined by experiment. As an approximate indication of whether gas-film or liquid-film resistance is likely to predominate, Wiegand (128) proposed the dimensionless criterion

$$\frac{1}{m} \frac{k_L a}{k_G a} \frac{C_{av}}{P}$$

where m = slope of equilibrium line in y/x coordinates,

$k_L a$, $k_G a$ = volume-basis mass-transfer coefficients for liquid and gas films, respectively,

C_{av} = average concentration of soluble gas in liquid, lb moles/ft;

P = total pressure, atm

This is in effect the ratio of the slope of the tie line to that of the equilibrium line in the phase diagram. Wiegand states that where the value of the ratio exceeds 10 gas-film control is likely; where it is below 0.1, liquid-film control is likely. At intermediate values, there will probably be mixed control. Morris and Jackson (122) have proposed a simpler criterion, for which it is not necessary to know the mass-transfer coefficients,

$$\frac{\rho_s}{HP}$$

where ρ_s = density of soluble gas at actual temperature and pressure of gaseous mixture, lb/ft^3, and H = solubility factor (= Henry's-law constant where Henry's law applies), lb/ft^3-atm.

When ρ_s/HP is less than 0.0005, gas-film control is likely; when it is more than 0.2, liquid-film control is likely; at intermediate values, mixed control is to be expected.

Liquid-film Control. Sherwood and Holloway (126) investigated the desorption of CO_2, oxygen and hydrogen from water, and found that the HTU varies as $d^{0.32}L^{0.24}$. Scheibel and Othmer (125), working with ketones and water, found the mass-transfer coefficient $k_L a$ to be independent of packing diameter and proportional to $L^{0.8}$, whence HTU varied as $L^{0.2}$. According to Sherwood and Pigford (127), HTU is proportional to L^n, where n varies from 0.22 to 0.46 according to the packing. Their tabulated figures, based partly on unpublished thesis results, suggest that, for packing diameters of 1 in. and over and at constant liquid and gas rates, HTU for ring packings increases approximately as $d^{0.25}$, whereas for Berl saddles it actually decreases. (The relation for rings is the safer for scaling up and is assumed for the purpose of Table 13-1). Morris and Jackson (122) state that k_1, the liquid-film coefficient (area basis), is proportional to $(L/a)^n$, where n varies from 0.75 to 0.95 according to the packing, whence for geometrically similar packings HTU$_L$ varies as $(Ld)^{1-n}$.

It is generally agreed that the gas velocity has little influence upon the liquid-film coefficient until near the flooding point.

Gas-film Control. Pratt (124) has shown that, whereas liquid-film volume-basis coefficients are little influenced by incomplete wetting of the packing at low liquor rates, gas-film coefficients are markedly reduced through this cause. Hence, the HTU of a packed tower increases for lower liquor rates and smaller packing diameters (larger specific surface). Sherwood and Holloway (126) concluded that for ammonia absorption HTU_G is proportional to $G^{0.5}/L^{0.4}$. The effect of packing diameter was not apparent. Scheibel and Othmer (125) found that $k_G a$ varies as $G^{0.8}/d$, whence HTU_G is proportional to $G^{0.2}d$. Sherwood and Pigford (127) give an equation according to which HTU_G varies as $G^{0.31}/L^{0.33}$, with no term for packing diameter. On the other hand, their Table XIX seems to show that, at constant liquid and gas rates and for diameters of 1 in. and over, the HTU_G varies as $d^{0.9}$ for Berl saddles and as $d^{0.7}$ for Raschig rings. (The former figure is the safer for scaling up and is assumed for the purpose of Table 13-1.) According to Morris and Jackson (122), k_g varies as $G^{0.75}$ and is almost independent of packing diameter, whence HTU_G is proportional to $G^{0.25}d$.

Evaporation

Gamson, Thodos, and Hougen (85) and Wilke and Hougen (91) investigated the rate of evaporation of water from packed solid shapes in an air stream. They found that the gas-film coefficient (area basis) k_g is proportional to G^{1-n}/d^n, where n is 0.41 for turbulent gas flow, 0.51 in the transition region, and 1.0 for streamline flow. Hence, HTU_g was proportional to $G^n d^{n+1}$, that is, for turbulent flow, to $G^{0.41}d^{1.41}$.

The evaporation of water into an air stream is necessarily subject to gas-film control, and yet the variation on HTU_g with packing diameter appears to be substantially greater than for the absorption processes mentioned earlier. This suggests that in gas-absorption processes the liquid-film resistance is seldom entirely negligible.

Liquid Extraction

Disperse-film Control. Colburn and Welsh (176), using isobutanol and water, found that HTU_d, the film HTU for the disperse phase, was practically independent of either flow rate. The packing diameter was not varied. Pratt and his coworkers (179, 183) investigated the transfer of acetone and acetaldehyde from various solvents to water and confirmed the findings of Colburn and Welsh. Lewis, Jones, and

Pratt (182) found that the droplet size of the disperse phase was constant for packing diameters of $\frac{1}{2}$ in. and upward.

Continuous-film Control. Colburn and Welsh (176) found that the film HTU for the continuous phase HTU_c varies as $(L_c/L_d)^{0.75}$, where L_c and L_d are the superficial mass velocities of the continuous and disperse phases, respectively. Laddha and Smith (181) found it to vary as $(L_c/L_d)^n$, where n varied from 0.54 to 0.96 according to the physical properties of the fluids. Pratt and his coworkers (179) state that HTU_c varies directly as L_c/L_d.

The literature on HTU in liquid extraction is mostly derived from towers of relatively small sizes, and there appear to be no published figures on the effect of large variations in either packing or tower diameter. Baron has stated that a 10-ft-diameter industrial extraction tower may have an HTU 15 to 20 times that found in a laboratory column a few inches in diameter. He attributes the relative inefficiency of the large tower to back mixing (175a). Pratt, in the course of recent unpublished work on the butyl acetate–acetone–water system, finds that the effects of changes in both packing and tower diameter depend very much on whether the direction of mass transfer is from the disperse to the continuous phase, or vice versa. His interim conclusions are as follows, (182a):

Disperse→continuous-phase transfer. The over-all HTU increases somewhat with tower diameter and markedly with packing diameter.

Continuous→disperse-phase transfer. The HTU increases with packing diameter, though less markedly than for disperse→continuous-phase transfer. It appears to be unaffected by changes in tower diameter provided that the tower diameter exceeds eight packing diameters.

Gas Absorption—Theoretical Relations

By making certain assumptions, it is possible to derive theoretical expressions for the effect on mass-transfer coefficient and HETP (or HTU) of fluid velocity and linear dimensions for the limiting cases of liquid-film controlling and gas-film controlling.

Liquid-film Control. Emmert and Pigford (121) have examined the theory of liquid-film control in wetted-wall towers. Assuming that the liquid film is in streamline flow and not appreciably disturbed by the gas flow and that the time of contact is long enough for the molecules of soluble gas to diffuse right through the film, they showed that the area-basis liquid-film mass-transfer coefficient k_l should be inversely proportional to the liquid-film thickness δ. For short contact times in which the molecules of solute penetrate only a small distance into the liquid film, k_l should be inversely proportional to $\sqrt{\delta}$.

Applying these relations to a packed tower, and assuming further that the packing is completely wetted by the liquid, δ is proportional to $\sqrt[3]{L/a}$, where a is the packing surface per unit volume [Eq. (13-1a)]. In the case of a long contact time,

$$k_l \propto \delta^{-1} \propto \left(\frac{a}{L}\right)^{\frac{1}{3}}$$

$$k_l a \propto \left(\frac{a^4}{L}\right)^{\frac{1}{3}} \propto \left(\frac{1}{Ld^4}\right)^{\frac{1}{3}}$$

where d is the packing diameter.   Hence,

$$\text{HTU}_l \propto \frac{L}{k_L a} \propto (Ld)^{\frac{4}{3}} \qquad (13\text{-}11a)$$

For the case of a short contact time,

$$k_l \propto \delta^{-\frac{1}{2}} \propto \left(\frac{a}{L}\right)^{\frac{1}{6}}$$

$$k_l a \propto \left(\frac{a^7}{L}\right)^{\frac{1}{6}} \propto \left(\frac{1}{Ld^7}\right)^{\frac{1}{6}}$$

$$\text{HTU}_l \propto \frac{L}{k_l a} \propto (Ld)^{\frac{7}{6}} \qquad (13\text{-}11b)$$

Hence, under otherwise constant conditions, the height of packing for a given separation should theoretically be proportional to somewhere between $d^{1.17}$ and $d^{1.33}$.   The experimental value found by Sherwood and Holloway (126) was only $d^{0.32}$.   This great difference between practice and theory may be attributed partly to the effect of gas flow in rippling the surface of the liquid film and creating additional surface that is not a function of the packing dimensions and partly to the fact that packings are seldom completely wetted, so that an increase in packing diameter though reducing the total surface increases the percentage wetted.

Gas-film Control.   Assuming that the flow of gas through the packing is turbulent and that the Prandtl group can be taken as constant, the regular equation for mass transfer by forced convection may be written*

$$\frac{k_g d}{D} = C \left(\frac{Gd}{\mu g}\right)^x \qquad (13\text{-}12)$$

* See Eq. (8-2).

where D = diffusion coefficient and x = Reynolds index, which for ring packings may be taken as 0.60

Hence, for homologous systems,

$$k_g \propto \frac{G^{0.6}}{d^{0.4}}$$

$$k_g a \propto \frac{G^{0.6}}{d^{1.4}}$$

$$\text{HTU}_g \propto \frac{G}{k_g a} \propto G^{0.4} d^{1.4} \qquad (13\text{-}13)$$

Under otherwise constant conditions, the height of packing for a given separation might be expected to vary as $d^{1.4}$. The correlations of Scheibel and Othmer (125) and Morris and Jackson (122) give a value of $d^{1.0}$, Gamson, Thodos, and Hougen (85) give $d^{1.41}$. Agreement with the theoretical figure is better than in the case of liquid-film control.

For perfect isothermal streamline flow, k_g is independent of the gas rate G. Hence,

$$k_g a \propto \frac{1}{d}$$

$$\text{HTU}_g \propto \frac{G}{k_g a} \propto Gd \qquad (13\text{-}14a)$$

In beds of broken solids and random packings, the transition from streamline to turbulent flow is much more gradual and extends over a wider range of velocities than in pipes. It is probable that the gas flow is seldom perfectly streamline even at low Reynolds numbers on account of the irregularities of the gas passages. In granular beds at values of dG/μ below 350, Wilke and Hougen (91) found that k_g was proportional to $\sqrt{G}$. Assuming this value, the above relation becomes

$$\text{HTU}_g \propto G^{0.5} d \qquad (13\text{-}14b)$$

Streamline or near-streamline gas flow is more likely in stacked-ring packings, since these tend to damp out eddies, after the manner of straightening vanes in a pipe. For this reason, the critical Reynolds number for stacked rings is about five times that for random packings.

The authors are not aware of any experimental determinations of the effect of d on HTU_g under streamline conditions with which the above theoretical relation could be compared.

Empirical Rules

Owing to the inherently mixed fluid regime in packed towers, it is not possible to derive scale equations from model theory alone; rules for scaling up the performance of experimental towers must be largely empirical and based on the correlations employed in tower design. The scaling-up method will, however, eliminate or greatly reduce errors due to shape factors or physical properties of the fluids.

The first design variable to be fixed by experiment will be the liquid/gas or disperse/continuous-phase ratio. It will be assumed that the optimum L/G ratio has been determined on the pilot scale and that the same ratio is to be employed on the large scale.

Packed towers are generally designed to operate at a given percentage of the flooding or loading velocities, since as the flooding point is approached the HETP or HTU diminishes. Therefore, it is recommended that model and prototype towers should be operated at equal percentages of their flooding velocities. Under otherwise constant conditions, the flooding velocity varies as the square root of the packing diameter; so the relation for scaling up fluid velocities in Eq. (13-5a). This relation assumes that packing elements are geometrically similar in themselves and in their method of packing. Commercial packings of different sizes are not always geometrically similar; for example, the thickness of rings may not vary in proportion to their diameter. Further, the voidage fraction of a given packing depends partly on the method of packing. Both the flooding velocity and the pressure drop in the tower are very much influenced by the voidage fraction; hence, for the purpose of scaling up, the "packing diameter" should be taken as the hydraulic mean diameter of the free space. Hence,

$$L \propto G \propto d^{1/2} \propto \left(\frac{4\epsilon}{a}\right)^{1/2} \tag{13-15}$$

ϵ can be determined experimentally or, in the case of Raschig rings, estimated by the empirical equation of Lobo, Friend, Hashmall, and Zenz (220):

For rings dumped in a tower dry:

$$\epsilon = 1.046 - 0.658M \tag{13-16}$$

For rings dropped into a tower filled with water:

$$\epsilon = 1.029 - 0.591M \tag{13-17}$$

For rings dropped into a tower filled with water and shaken until the packing assumed its most dense arrangement:

$$\epsilon = 1.009 - 0.626M \tag{13-18}$$

where
$$M = \frac{(1 - d_i/d_o)^2}{(hd_o)^{0.017}} \tag{13-18a}$$

where h = ring height, in.

d_o = outside diameter of ring, in.

d_i = inside diameter of ring, in.

The average deviation of experimental measurements from these formulas was ± 2.5 to 2.7 per cent.

The relative gas and liquid velocities defined by Eq. (13-15) represent the nearest approach to corresponding velocities that is possible under the mixed regime obtaining in a packed tower. At these velocities, the gaseous pressure drop per foot height of packing (Δp) is approximately constant. It is exactly constant at the loading point as defined by Morris and Jackson (122), when Δp is proportional to G^2/d. At lower fluid velocities, Δp become proportional to $G^{1.8}/d^{1.2}$, whence for the relation of velocities to packing diameter defined by Eq. (13-15) Δp varies $1/d^{0.3}$.

There are two practical limitations to the scaling up of packing diameter. First, if the packing diameter in the pilot-scale tower is too small, then surface-tension effects are exaggerated and the relation between linear dimension and performance is not the same as for larger rings. Data published by Sherwood and Pigford on gas absorption (127) and Lewis, Jones, and Pratt on liquid extraction (182) suggest that, if the performance of an experimental tower is to be scaled up quantitatively, its packing should certainly be not less than $\frac{1}{2}$ in. diameter, and preferably not less than $\frac{3}{4}$ in. If a tower-diameter/-packing-diameter ratio of 8:1 be accepted as the minimum permissible without excessive wall effect, then the experimental tower should be not less than 4 in. diameter for $\frac{1}{2}$-in. packing or 6 in. diameter for $\frac{3}{4}$-in. packing. For ring packings, Baker, Chilton, and Vernon (212) recommend a ratio of not less than 15:1.

The second limitation on scaling up is that both model and prototype packings must be adequately wetted without being flooded. Morris and Jackson (122) quote the following minimum wetting rates:

For rings of diameter greater than 3 in. and grids of pitch greater then 2 in.:

$$\frac{L}{a} = 1.3 \text{ ft}^3/\text{hr-ft} \qquad (13\text{-}19a)$$

For all other packings: $\quad \dfrac{L}{a} = 0.85 \text{ ft}^3/\text{hr-ft} \qquad (13\text{-}19b)$

Since it has been decided that L shall vary as $\sqrt{d}$, and a varies as $1/d$,

$$\frac{L}{a} \propto d^{1.5} \qquad (13\text{-}20)$$

If the scale ratio is large, then either the model packing will be incompletely wetted or the prototype packing will be overloaded with liquid. Morris and Jackson recommend designing for $2\frac{1}{2}$ times the minimum wetting rate, although higher values are permissible.

It is therefore recommended that the wetting rate $L/\rho_l a$ in the model tower should be held at the minimum value given by Eq. (13-19a) or (13-19b) or slightly below it (correction factors for incomplete wetting are given by Morris and Jackson, but the effects are slight up to 10 per cent below the specified minimum wetting rate). The scale ratio for packing diameter should not exceed 3, or preferably 2, which last would give a wetting rate in the prototype 2.84 times that in the model. Assuming a constant ratio of tower to packing diameter (D_b/d), a constant L/G ratio, and fluid velocities varying as Nd, then, if q be the total throughput per hour in geometrically similar towers,

$$q \propto Gd^2 \propto d^{2.5} \tag{13-21a}$$
or
$$d \propto q^{0.4} \tag{13-21b}$$

Thus, for a scale ratio of 2, the prototype tower would have 5.7 times the throughput of the model and, for a scale ratio of 3, 16.2 times the throughput. If a greater throughput than this is required on the large scale, then D/d for the prototype will have to be increased and the small-scale tower will become a model element. It is probably better to limit the scale ratio to 2 and vary the D/d ratio than to have a scale ratio of 3 or more with D/d constant.

The scaling up of the packed height depends upon the type of operation and whether the gas or liquid film is controlling. Assuming L/G constant and $L \propto G \propto \sqrt{d}$, the HTU and total packed height for a given separation will vary as d^n, where n is an empirical exponent. Table 13-1 shows values of n derived from various sources. A study of these figures and the literature from which they were derived suggests the following empirical rules for scaling up packed height and HTU under the liquid-and gas-flow conditions specified above.

In gas absorption with liquid-film control, the experimental evidence suggests that within the limits previously mentioned pilot-plant results can safely be scaled up according to the relation

$$\text{HTU}_l \propto d^{0.5} \tag{13-22}$$

In gas absorption with gas-film control, most of the experimental results indicate that HTU_g varies approximately as d. However, the results for the evaporation of water which is also gas-film-controlled, together with the theoretical relation, suggest that for safety the relation should be

$$HTU_g \propto d^{1.5} \qquad (13\text{-}23)$$

This assumes turbulent flow of gas.

For mixed gas- and liquid-film control, no rule can be given. As an approximation, some value of n might be selected between 0.5 and 1.5 according to the estimated relative influence of the liquid- and gas-flow

TABLE 13-1. SCALE-UP RELATIONS FOR HTU IN PACKED TOWERS

HTU $\propto d^n$, where d = equivalent packing diameter = $4\epsilon/a$. L/G ratio (or v_d/v_c for liquid/liquid) assumed constant, tower/packing-diameter ratio constant, L (or v_d) varied as $d^{1/2}$.

Operation	Controlling film	n	Date	Reference	Remarks
Distillation........	.....	1.0	1922	89	
		1.0	1948	88	
		0.85	1950	87	
		1.1	1950	90	
		0.25	1951	86	
Gas absorption....	Liquid	0.44	1940	126	
		0.1	1944	125	
		0.42	1952	127	
		0.5	1953	122	
	Gas	1.1	1944	125	
		0.89	1952	127	
		1.125	1953	122	
		1.5	....	...	Theory streamline flow
		1.6	....	...	Theory turbulent flow
Evaporation.......	Gas	1.61	1943	85	Turbulent flow
		1.77	1945	91	Transition region
		2.50	1943	85	Streamline flow

films, respectively, on the total resistance to mass transfer. Probably the safest procedure is to conduct experiments on a full-scale element of the prototype tower.

The experimental results for distillation indicate values of n intermediate between those for gas-film and liquid-film control, which suggests that the regime in distillation is one of mixed-film control. It is customary to assume the relation

$$HTU \text{ (or HETP)} \propto d \qquad (13\text{-}24)$$

In liquid extraction, the problem of scaling up is particularly difficult. Large variations in HTU with both packing and tower diameter are reported to occur in some circumstances, with rather slight variations in other circumstances, but there is as yet no systematic correlation. In the absence of any other information, the authors would very tentatively propose the following relation:

$$HTU \text{ (or HETP)} \propto d^{0.75}$$
$$\propto D^{0.25} \qquad (13\text{-}25)$$

Summary

The mixed gravity–viscosity–surface-tension-controlled regime that exists in countercurrent packed towers makes it impossible to be exact either in designing them from laboratory data or in scaling up from pilot units. A direct scale up may nevertheless avoid some of the sources of error inherent in design from generalized correlations. The following empirical rules are suggested:

1. Packing diameter in the small-scale tower to be certainly not less than ½ in. and preferably not less than ¾ in.

2. Diameter of small-scale tower to be not less than 8 times the packing diameter, or for ring packings that are wetted by the disperse phase preferably not less than 15 times the packing diameter.

3. Wetting rate in the small-scale tower to be the minimum as given by Equation (13-19b) or slightly less.

4. Liquid/gas or disperse/continuous-phase ratios in the small- and large-scale towers to be identical.

5. Scale ratio of packing diameters to be certainly not more than 3 and preferably not more than 2.

6. Both fluid velocities to be proportional to the square root of the hydraulic mean diameter of the packing [Eq. (13-15)].

7. Scale-up of HTU or HETP to vary according to the operation and control:

Gas absorption, liquid-film control..........	Eq. (13-22)
Gas absorption, gas-film control ⎱	Eq. (13-23)
Evaporation ⎰	
Distillation...............................	Eq. (13-24)
Liquid extraction.........................	Eq. (13-25)

8. In gas-liquid or vapor-liquid towers under the above conditions, the pressure drop per foot height of packing will be the same in model and prototype if both are operated at the loading point, or may vary as $1/d^{0.3}$ if substantially below it.

Examples. 1. An experimental distillation column of 8 in. inside diameter is packed with ½-in. rings for a height of 6 ft. It is found to effect a given separation at a reflux ratio of 2:1, yielding 70 lb/hr of distillate, at which rate the column is operating near the flooding point. The pressure drop across the packing is 5 in. water gauge. What will be the output of a column of 2 ft diameter packed with 1-in. rings, what height of packing will be required to give an equivalent separation at the same reflux ratio, and what will be the pressure drop?

To maintain the same ratio of operating to flooding rate in the larger column, the vapor velocity must be increased in proportion to the square root of the ring diameter [Eq. (13-5a)]. Hence, the ratio of vapor throughputs is given by

$$\frac{24^2 \times \sqrt{2}}{8^2} = 12.72$$

The output of the larger column at the same reflux ratio will therefore be $70 \times 13.72 = 890$ lb/hr of distillate.

In accordance with Eq. (13-24), the packed height should be twice that in the experimental column = 12 ft.

The pressure drop per foot height in both columns will be approximately the same, giving a total pressure drop of 10 in. water gauge across the larger column.

2. A tower of 6 in. inside diameter packed with $\frac{3}{4}$-in. porcelain rings is used to scrub traces of acid vapors from the residual gases issuing from a pilot plant. A low pressure drop across the packing is essential. It is found experimentally that with a water flow of 12 lb/hr and a packed height of 3 ft 6 in. the tower will satisfactorily scrub 320 ft³/hr of gases (at NTP). The pressure drop across the packing is then 0.6 in. water gauge. Specify the dimensions of a tower to scrub 10,000 ft³ of gases per hour, and calculate the pressure drop.

For the small tower:

$$a = 72 \text{ ft}^2/\text{ft}^3$$
$$L = 61.3 \text{ lb/ft}^2\text{-hr}$$
$$\Delta p = 0.17 \text{ in. water gauge/ft height}$$

From Eq. (13-19b), it may be checked that 61.3 is about the minimum wetting velocity for this packing.

For the large tower, 2-in. stoneware rings are chosen, giving a linear scale ratio of $d = 2/0.75 = 2.67$. Since the vapors to be absorbed are dilute, it is safe to assume gas-film control. From Eq. (13-23), the height of packing in the large tower should be

$$3.5 \times 2.67^{1.5} = 15.2 \text{ ft}$$

For 2-in. stoneware rings, the specific surface $a = 29$ ft²/ft³. If the large tower is to operate at $2\frac{1}{2}$ times the minimum wetting rate, then from Eq. (13-19b) the liquid rate will be

$$L = 2.5 \times 0.85 \times 29 = 61.5 \text{ lb/ft}^2\text{-hr}$$

Preserving the same gas-liquid ratio as in the small tower, the volumetric gas rate will be

$$\frac{320 \times 61.5}{12} = 1,640 \text{ ft}^3/\text{ft}^2\text{-hr}$$

The cross-section required for 10,000 ft³/hr is thus 6.1 ft², corresponding to an inside diameter of 2.79 ft.

Expressing Eq. (13-7a) in ratio form (boldface type denoting value for prototype ÷ value for model),

$$\Delta p = \frac{G^{1.85}}{d^{1.15}}$$

where $d = 2.67$ and $G = (1,640 \times 0.196)/320$, whence

$$\Delta p = 0.32$$

Δp for the prototype tower will therefore be $0.17 \times 0.32 = 0.055$ in. water gauge per foot height, giving a total pressure drop of $0.055 \times 15.2 = 0.84$ in. water gauge

The required figures for the full-sized tower are therefore:

Inside diameter . . . .   2.79 ft
Packing . . . .   2-in. stoneware rings
Packed height . . . .   15.2 ft
Pressure drop . . . .   0.84 in. water gauge

SYMBOLS IN CHAPTER 13

A = const in Eq. (13-1a)
a = specific surface of packing, ft^2/ft^3
a = const in Eq. (13-10)
B = friction factor in Eq. (13-6)
b = const in Eq. (13-10)
C = concentration of soluble gas in the liquid
D = tower diameter, ft
d = packing-element diameter, ft
d_i = inside diameter of ring, in., Eq. (13-18a)
d_o = outside diameter of ring, in., Eq. (13-18a)
G = gas mass velocity based on empty tower, lb/ft^2-hr
g = acceleration of gravity
H = solubility factor = Henry's-law constant where Henry's law applies, lb/ft^3-atm
h = height of ring, in., Eq. (13-18a)
K = over-all mass-transfer coefficient, area basis
k = film coefficient of mass transfer, area basis
L = liquid mass velocity based on empty tower, lb/ft^2-hr
M = a quantity defined by Eq. (13-8a)
m = slope of equilibrium line in y/x coordinates
n = an exponent
P = total pressure, atm
Δp = pressure drop in gas phase per foot height of packing
v = linear velocity
w = total flow of liquid, lb/hr
Z = packed height of tower, ft
δ = film thickness, ft
ϵ = voidage fraction of irrigated packing
μ = viscosity
ρ = density

Subscripts

av = average
c = pertaining to the continuous phase
d = pertaining to the disperse phase
g = gas basis, pertaining to the gas
l = liquid basis, pertaining to the liquid
s = pertaining to the soluble gas

CHAPTER 14

MIXING EQUIPMENT

The principal applications of mixing equipment in the process industries are:

1. Blending of solid powders or pastes
2. Suspension of solids in liquids
3. Dispersion or emulsification of immiscible liquids
4. Solution of solids, liquids, or gases
5. Promotion of chemical reactions

Any of these operations may be accompanied by heating or cooling in the mixer.

There is little quantitative information about the mixing of solids and pastes and none on which scale relations could be based. This chapter is therefore confined to operations involving mixing in liquid media, the standard equipment for this purpose being a vessel furnished with a rotating stirrer. The stirred vessel as a heat-transfer apparatus is considered in Chap. 12 and as a chemical reactor in Chap. 15. The present chapter is concerned with physical mixing alone or accompanied by heat or mass transfer. It is assumed that a satisfactory degree of mixing has been achieved experimentally on the pilot scale and is required to be reproduced in a full-sized unit.

The fluid-flow pattern in a mixer is complex and influenced largely by the geometry of the system. The differential equations of flow can be written down, but they cannot be generally integrated. For this reason, considerable use has been made of models in the study of mixing. In a classical series of papers (191 to 197), Hixson and his coworkers investigated the behavior of geometrically similar mixers of different sizes. Other model studies have been carried out by Buche (188) and Brothman and Kaplan (187). Rushton (207 to 209) has set out the general principles involved in extrapolating the performance of a model mixer to higher Reynolds numbers on the large scale. Hixson and Baum (191, 192) and Chilton, Drew, and Jebens (148) found, respectively, that mass- and heat-transfer coefficients in geometrically similar mixers were equal when the peripheral speeds of the agitators were equal.   Buche (188), Brothman and Kaplan (187), and Miller and Mann (204), on the other hand, concluded that small and large mixers gave equal mass-transfer coefficients and mixing rates at equal power inputs per unit volume.   Rushton (207) showed that either statement could be correct under appropriate conditions but that both were special cases of a more general set of scale relations involving the parameter that is here referred to as the Reynolds index. These relations are in principle applicable over a wider field than mixing, and they have already been discussed in Chap. 8.   Vermeulen, Williams, and Langlois (211) used an optical method to determine the specific interface in immiscible fluid systems.

The generalized dimensionless equations for fluid motion and heat and mass transfer that were derived in Chap. 4 and 5 and applied to scale relations in Chap. 7 are applicable to power consumption and heat and mass transfer in mixers.   It is customary to replace the mean fluid velocity v by the peripheral speed of the agitator, which is proportional to Nd, where d = diameter of agitator or length normal to axis of rotation and N = angular velocity of agitator, rps or revolutions per hour according to the units employed.   The linear dimension L in the generalized groups is replaced by the agitator diameter d.   Finally, the pressure coefficient is expressed in terms of the power consumption P and is called the *power number*.   The resulting generalized dimensionless equation for fluid flow in a mixer is

$$\frac{Pg_c}{\rho N^3 d^5} = \phi\left(\frac{Nd^2\rho}{\mu}, \frac{N^2 d}{g}\right) \tag{14-1}$$

Power number   Rey-   Froude
nolds   num-
num-   ber
ber

This is strictly analogous to Eq. (5-25).   Where there are two immiscible liquids, the Weber number enters and the relation becomes

$$\frac{Pg_c}{\rho N^3 d^5} = \phi\left(\frac{Nd^2\rho}{\mu}, \frac{N^2 d}{g}, \frac{N^2 d^3 \rho}{\sigma}\right) \tag{14-2}$$

This is analogous to Eq. (4-3).

It is known that the effects of the Reynolds and Froude numbers in the fluid-flow equation can with fair accuracy be represented by power functions of these numbers, although the exponents are not constant over the whole range of velocities.   The Weber number enters where the system contains immiscible fluids, either liquid or gaseous.   Earlier investigators concluded that for such systems the degree of dispersion of the two phases was a function of the horsepower input per unit volume (119, 204). Vermeulen, Williams, and Langlois (211) measured the mean droplet or bubble diameter in liquid-liquid and gas-liquid dispersions by an optical method and found that it was a power function of the Weber number alone for liquid-liquid systems and of the Weber and Reynolds numbers for gas-liquid systems.

Where there is a single fluid phase, the general equation for fluid motion in a mixer is given by Rushton, Costich, and Everett (209),

$$\frac{Pg_c}{\rho N^3 d^5} = C \left(\frac{Nd^2\rho}{\mu}\right)^m \left(\frac{N^2 d}{g}\right)^n \tag{14-3}$$

The exponent m varies from -1.0 in the streamline region to zero for fully developed turbulence in baffled mixers. The fluid motion is streamline at Reynolds numbers below 10 and fully turbulent at values of 100 to 1,000 depending on the geometry of the system.

The exponent n is zero in baffled mixers: in unbaffled mixers, it was found by Rushton et al. to be equal to

$$\frac{a - \log (Nd^2\rho/\mu)}{b}$$

where the constants a and b depend upon the geometry.

In an unbaffled mixer, the power number is thus influenced by both the Reynolds number and the Froude number. The Reynolds number determines the viscosity-controlled turbulence pattern within the fluid, while the Froude number determines the vortex profile and the toroidal circulation in a vertical plane, which are controlled by gravity. From Eq. (14-1) or (14-3), it is evident that for homologous unbaffled-mixer systems similarity is impossible between equipment of different sizes since, as has been noted earlier, the speed relations for equality of Reynolds and Froude numbers are incompatible. Rushton (207) has pointed out, however, that, if what is aimed at is simply a model experiment to determine the total flow pattern and power consumption in an unbaffled mixer, this can be achieved by employing a liquid of lower kinematic viscosity in the model than in the prototype. Using, as before, boldface type to denote ratios of quantities in prototype and model ($\mathbf{d} = d$ prototype/d model, $\mathbf{\nu} = \nu$ prototype/ν model, etc.), simultaneous equality of Reynolds and Froude numbers can be attained in geometrically similar unbaffled mixers if the agitator diameters and speeds are related to the ratio of kinematic viscosities as follows:

$$\mathbf{d} = \mathbf{\nu}^{2/3} \tag{14-3a}$$
$$\mathbf{N} = \mathbf{\nu}^{-1/3} \tag{14-3b}$$

Under these conditions, the power numbers in model and prototype are equal, whence the ratio of power consumptions is given by

$$\mathbf{P} = \rho \mathbf{N}^3 \mathbf{d}^5 = \rho v^{7/3} \tag{14-3c}$$

Mercury at 20°C has a kinematic viscosity $1/8.9$ times that of water, so that the power characteristics of an unbaffled mixer for aqueous solutions could be predicted from a geometrically similar model to a scale ratio of $8.9^{2/3} = 4.3$ containing mercury. Strict geometrical similarity extends to surface irregularities; therefore, the submerged surfaces of such a model should be polished. The power input to the model may be calculated from the agitator speed multiplied by torque, the latter being determined by mounting the mixer vessel on a torsion balance. Under the conditions of Eqs. (14-3a) to (14-3c), the vortex patterns and profiles in model and prototype will be geometrically similar.

In a mixer which forms part of a pilot plant, the liquid is the same on the small and the large scale, and similarity is impossible unless the effect of the Froude number is eliminated. This may be achieved by providing submerged baffles of sufficient area to suppress vortex formation entirely. A mixer so fitted is said to be fully baffled. Baffles increase the power consumption and (except at low Reynolds numbers) the mixing efficiency, so that they are generally employed where good mixing is important. It is only when vortex formation is suppressed that homologous mixing systems can be quantitatively scaled up.

Instead of providing baffles, it is possible to suppress vortex formation by offsetting the stirrer from the center line of the mixing vessel. Rushton and Oldshue (210) have pointed out that for the complete suppression of swirl an offset stirrer should be of the axial-flow (propeller) type and slightly inclined in such a manner that the axial component of flow tends to oppose the swirl induced by the rotation of the stirrer (198). An offset mixer in which swirl is virtually suppressed is equivalent to a fully baffled mixer in being independent of the Froude number.

The power equation for a fully baffled mixer is

$$\frac{P g_c}{\rho N^3 d^5} = C \left(\frac{N d^2 \rho}{\mu} \right)^m \tag{14-4}$$

To bring the power equation into a form comparable with the standard heat- and mass-transfer equations, Olney and Carlson (205) multiply both sides of Eq. (14-4) by the Reynolds number. Replacing P/d^3 in their equation by the power input per unit volume $= \pi$, this gives

$$\frac{\pi g_c}{N^2 \mu} = C_1 \left(\frac{Nd^2 \rho}{\mu}\right)^{x_f}$$ (14-5)

where x_f, the Reynolds index for fluid friction or momentum transfer, $= 1 + m$. The group on the left is a modified form of the group termed by Olney and Carlson the *power function*, and it represents the ratio of applied torque to viscous forces. Retaining the agitator diameter d as the characteristic length for the system, and assuming the Prandtl and Schmidt numbers to be constant, the corresponding heat- and mass-transfer equations for homologous mixing systems are

$$\frac{hd}{k} = C_2 \left(\frac{Nd^2 \rho}{\mu}\right)^{x_h}$$ (14-6)

$$\frac{Kd}{D} = C_3 \left(\frac{Nd^2 \rho}{\mu}\right)^{x_m}$$ (14-7)

where x_h and x_m are the Reynolds indices for heat and mass transfer, respectively. Typical values are given in Table 8-1.

Equations (14-5) to (14-7) apply only where the transfer surface is fixed or at least calculable. Thus, the mass-transfer equation is applicable only to the solution or crystallization of solid particles. Immiscible liquids and liquid-gas mixtures are subject to the influence of the Weber group.

As in other fluid-flow operations, it is seldom feasible to compare homologous systems at equal Reynolds numbers, except for small scale ratios, owing to the excessive stirrer speed and power consumption called for in the model. Hence, in scaling up the performance of a baffled mixer, it is necessary to employ the methods of extrapolation proposed by Rushton (207), which in their general aspects have already been discussed in Chap. 8. For scaling up power consumption, it is essential to know x_f and, for heat and mass transfer, x_h and x_m, respectively. These indices are somewhat dependent on the geometry of the system, and the most reliable scaling-up procedure is to determine them over a wide range of Reynolds numbers in a model mixer. Where this is not possible, appropriate values must be selected from the literature.

Table 8-1 shows Reynolds indices for heat and mass transfer as determined by a number of workers in various geometrical systems. For stirred vessels in which the area of the heat- or mass-transfer surface is independent of the fluid motion, a value of 0.6 may be taken for both x_h and x_m. For x_f, the Reynolds index for momentum transfer, Olney and Carlson (205) suggested that as an approximation a constant value of 0.85 might be applied to both baffled and unbaffled mixers over the whole range of turbulent flow. The actual value is influenced substantially by baffling and Reynolds number and to a lesser extent by the geometry of the agitator and mixing vessel. Martin (203) and Rushton, Costich, and Everett (209) have plotted power number against Reynolds number for their own experiments and those of other workers. From these plots, which together cover 27 different types of mixer investigated by 10 different sets of workers, the generalized curves of Fig. 14-1 have been derived. At Reynolds numbers

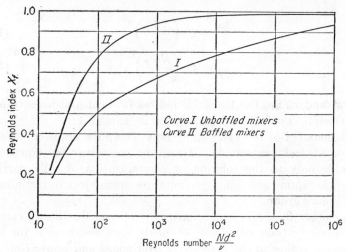

FIG. 14-1. Reynolds index for momentum transfer in mixers. [*J. J. Martin, Trans. AIChE,* **42**: 777 (1946); *J. H. Rushton, E. W. Costich, and Everett, Chem. Eng. Progr.*, **46**: 395, 467 (1950).]

below 10, the fluid motion in all mixers was streamline. Above 300, it was fully turbulent. Between 10 and 300 lay the transition region, which is considerably more spread out than in pipe flow. It will be seen from Fig. 14-1 that at Reynolds numbers above 300 Olney and Carlson's average value of 0.85 for x_f is roughly correct in unbaffled mixers, but in fully baffled mixers the average value of x_f is almost unity. For certain turbine mixers examined by Rushton et al. (209), x_f was slightly above unity at Reynolds numbers between 10^3 and 10^4.

For the dispersion of immiscible liquids or of gases in liquids, the work of Vermeulen, Williams, and Langlois (211) provides a quantitative basis. If s is the specific interface of the two-phase system, inversely proportional to the mean droplet or bubble diameter, then for geometrically similar mixers and equal volumetric proportions of the two phases the empirical equations of Vermeulen et al. may be written, for liquid-liquid systems,

$$sd \propto \left(\frac{N^2 d^3 \rho}{\sigma}\right)^{0.6} \qquad (14\text{-}8)$$

The mean deviation from the above relation in experiments with various pairs of liquids was 20 per cent. For homologous systems, the above relation reduces to

$$s \propto (N^3 d^2)^{0.4} \qquad (14\text{-}8a)$$

For gas-liquid systems, Vermeulen et al. found that

$$sd \propto \left(\frac{N^2 d^3 \rho}{\sigma}\right)\left(\frac{N d^2 \rho}{\mu}\right)^{0.5} \qquad (14\text{-}9)$$

Here the Weber number is divided by the square root of the Reynolds number. For homologous systems, the relation reduces to

$$s \propto (N^3 d^2)^{0.5} \qquad (14\text{-}9a)$$

If the proportions of the two phases present in the model and prototype are not the same, then an empirical correction factor has to be applied to the product sd in the above equations (211).

The rule for extrapolating mixer performance varies according as the object is to disperse a solid or immiscible liquid, on the one hand, or to effect heat or mass transfer, on the other. For preparing dispersions of solids in liquids, the published evidence suggests that the most suitable criterion is power input per unit volume: i.e., in homologous mixing systems at equal values of π the degree of dispersion is equal. Since it is only fully baffled or correctly offset mixers that can be scaled up on Reynolds index alone, the value of x_f may be taken as 1.0. Hence, from Eq. (14-5),

$$\pi = \frac{C_1}{g} N^3 d^2 \rho \qquad (14\text{-}10)$$

whence, for π constant,

$$\mathbf{N} = \mathbf{d}^{-\frac{2}{3}} \qquad (14\text{-}10a)$$
$$\mathbf{P} = \mathbf{d}^3 \qquad (14\text{-}10b)$$

These are the scale equations for the mixer as a disperser. At relative stirrer speeds corresponding to (14-10a), the relative power consumptions correspond to (14-10b).

In liquid-liquid and gas-liquid systems, true similarity requires that the Weber numbers shall be equal in model and prototype. N^2d^3 is then constant, and the scale equations are

$$N = d^{-\frac{2}{3}} \qquad (14\text{-}11a)$$

$$s = d^{-1} \qquad (14\text{-}11b)$$

This relation is not so impracticable as that of equal Reynolds numbers, but it still calls for an inconveniently high stirrer speed in the model, since the peripheral velocity is required to be inversely proportional to the square root of the linear scale. It is more convenient to extrapolate to the condition of equal specific interface in model and prototype by means of Eq. (14-8a) or (14-9a). For a constant value of s, both equations lead to the same scale equation, N^3d^2 constant, which reduces to Eq. (14-10a). It appears therefore that for immiscible liquids and gas-liquid systems also, equality of power input per unit volume gives equal degrees of dispersion.

For scaling up heat-transfer coefficients in mixers, the appropriate scale equations have been given in Chap. 12, Eq. (12-6)ff. In general, the ratio of stirrer speeds at which heat-transfer coefficients are equal is not the same as that at which power inputs per unit volume and specific interfaces are equal. Hence, in scaling up mixer performance where heat transfer is involved, it is advisable to utilize Eqs. (14-10a) and (14-10b) for stirrer speed and power and adjust the heat-transfer coefficient by the general scale-up equation (12-6), choosing an appropriate value of the Reynolds index.

Example. A heavy tar-oil emulsion is to be prepared batchwise in a steam-jacketed pan fitted with a stirrer. The procedure is to charge the pan with cold water, heat it to 60°C, using the stirrer to accelerate heat transfer, then run in the correct quantity of hot tar oil mixed with emulsifying agent from a separate tank. Pilot-scale experiments were conducted in a jacketed pan 10 in. in diameter having a propeller stirrer running at 1,500 rpm. The heating-up time for a batch was 2.4 min, and the measured power consumption was 0.004 hp. The full-sized unit is a geometrically similar jacketed pan 5 ft in diameter having a geometrically similar stirrer. What should the stirrer speed be to give a degree of dispersion equal to that obtained in the small unit, what will be the power consumption, and how long will a charge of water take to heat up?

The linear scale ratio $L = d = 60$ in./10 in. = 6.0.

From Eq. (14-10a), the ratio of stirrer speeds should be $1/6^{\frac{2}{3}} = 1/3.3$. The correct speed for the full-sized stirrer is therefore

$$\frac{1,500}{3.3} = 455 \text{ rpm}$$

From Eq. (14-10b), the power consumption of the full-sized unit will be

$$0.004 \times 6^3 = 0.86 \text{ hp}$$

The jacketed heating surfaces are geometrically similar and the model has no dummy surface; therefore, it has 6 times as much heating surface per unit volume as the large pan. Further, the heat-transfer coefficient in the large pan must be scaled down in accordance with Eq. (12-6).

The ratio of peripheral stirrer speeds $v = Nd = 6.0/3.3 = 1.82$.

From Eq. (12-6), taking the Reynolds index as 0.6,

$$h = \frac{1.82^{0.6}}{6.0^{0.4}} = \frac{1}{1.43}$$

Hence, the heating-up time in the full-sized pan will be

$$2.4 \times 6.0 \times 1.43 = 20.6 \text{ min}$$

SYMBOLS IN CHAPTER 14

a, b = const relating to the value of n in unbaffled mixers
C, C_1, C_2, C_3 = const
D = diffusion coefficient
d = diameter of stirrer
g = acceleration of gravity
g_c = Newton's-law conversion factor
h = film coefficient of heat transfer
K = film coefficient of mass transfer
k = thermal conductivity of fluid
L = linear dimension
m = exponent in Eq. (14-3)
N = rotational speed of stirrer
n = exponent in Eq. (14-3)
P = power consumption of stirrer
s = specific interface, area per unit volume
v = fluid velocity, or peripheral stirrer speed
x = Reynolds index
x_f = Reynolds index for fluid friction or momentum transfer $= 1 + m$
x_h = Reynolds index for heat transfer
x_m = Reynolds index for mass transfer
μ = fluid viscosity
ν = kinematic viscosity of fluid $= \mu/\rho$
π = power input to mixer per unit volume
ρ = fluid density
σ = interfacial tension

CHAPTER 15

CHEMICAL REACTORS

Chemical reactions are carried out industrially in vessels of various types according to the physical state of the reactants, temperature and pressure of the reaction, agitation and heat-transfer requirements, and whether the reaction is to be conducted batchwise or continuously. Batch liquid-phase reactions at atmospheric pressure are generally carried out in open or loosely covered pots provided with stirrers and jackets or coils for heating and cooling. For batch reactions under pressure or vacuum, the corresponding vessels are autoclaves and vacuum pans, respectively. Continuous liquid-phase reactions at atmospheric pressure may be conducted in a series of stirred pots through which the reactants are passed—up to 16 pots are used in some processes. For continuous reactions under pressure or vacuum, or where the fluid involved is a gas, closed reaction vessels are employed which vary in shape from short cylinders to long tubes and which may contain solid catalysts.

It is sometimes possible to design a full-scale reactor successfully from a thermodynamic and kinetic analysis of laboratory data alone and without any pilot-plant experimentation. This is especially so in the case of catalytic reactors, where a short tube packed with catalyst can be treated as a differential element of the full-sized reactor. Theoretical methods of design on this basis have been developed by Hurt (17), Hulburt (16), Dodd and Watson (12, 13), Brinkley (5), and others. Jones has described a useful graphical method of designing a series of continuous stirred-tank reactors from laboratory batch-reaction curves (19). The direct scaling up of pilot-plant reactors on similarity principles has been discussed by Damköhler (7, 8), Edgeworth Johnstone (18), and Bosworth (1 to 4).

Whether a chemical reactor should be designed directly from laboratory data or first transferred to the pilot scale depends upon a number of factors: the state of knowledge of the chemistry of the reaction (including side reactions), the desirability of starting with a "trickle" production to test the market, the urgency of reaching full-scale production, and the degree of risk that the financial authority is prepared to accept. Many chemical-manufacturing processes have operated successfully and earned profits long before their chemistry was fully understood, and there are circumstances in which it might be quicker and less costly to develop a pilot-plant reactor by empirical methods and scale it up directly rather than to undertake a complete elucidation of the reaction mechanism. In other cases, the omission of the pilot-plant stage will be justified, but where for any of the above reasons it is decided to construct a pilot unit, the question of scale relations has to be considered.

The performance of a chemical reactor may involve every one of the rate processes with which chemical engineers are normally concerned: fluid friction, heat transfer, mass transfer (mixing), and chemical change. The scale relations for the first three processes have already been dealt with, and in the present chapter it will be assumed that *the main object in scaling up a model reactor is to secure the same yield of product on the large scale as on the small.* As mentioned in Chap. 7, it is theoretically possible to have chemical similarity between systems involving two different reactions of the same order, but partly owing to the influence of side reactions on yield the practical scaling up of chemical reactions is limited to homologous systems.

The yield of a chemical reaction is governed by three factors: the equilibrium state, the reaction velocity, and the residence time in the reaction system. The thermodynamic relations for the equilibrium constant K and the velocity constant k are, respectively,

$$K \propto \epsilon^{-\Delta F/RT}$$
$$k \propto \epsilon^{-E/RT}*$$

where ΔF = free-energy change at standard activity and E = activation energy.

If these two constants are to have the same respective values on the large and the small scale, then the absolute temperatures of the two systems must be the same.

Equality of residence times presents no problem in the case of batch reactions, since the whole of the material is held in the reaction zone for a determined period, after which the reaction is stopped. The only chemical-kinetic problem in scaling up a batch reaction is where the reaction is positively or negatively catalyzed by the wall of the reaction vessel, in which case strict similarity calls for an equal ratio of surface to volume on the large scale, or, alternatively, the wall of the

* See Eq. (5-60).

small-scale vessel might be treated to render it partially inactive, as suggested in Chap. 8. Otherwise, equality of chemical yield does not require geometrical similarity of reaction vessels. It may nevertheless be desirable to have the vessels similar in order that heat-transfer rates and power consumption may be scaled up or extrapolated by the methods of Chaps. 12 and 14. With equal heat-transfer coefficients in the small and large reaction vessels, any fouling or corrosive effects on heating surfaces are likely to be similar in both.

The problem of residence time is a real one in continuous-flow reactors. There is no practicable means of securing "plug" or "piston" flow of fluid through a continuous reactor, whereby every particle of fluid remains in the reaction zone for the same time. Therefore, instead of a single residence time, there is a statistical distribution function. This may be represented in various ways, one of which is by means of Danckwerts's F diagram (189). Imagine a vessel of volume V into and out of which fluid is flowing at a steady rate. At a time $t = 0$, the color of the incoming fluid changes from, say, white to red. Let Q be the total volume of outflow from the time the change takes place, i.e., when $t = 0$, $Q = 0$. Then $F(t)$ would be the fraction of red fluid in the outflow after a time t. It is plotted against Q/V. For true plug flow, $F(t)$ rises vertically from zero to unity at $Q/V = 1.0$. Longitudinal mixing converts the vertical straight line into an ogive. Where there is perfect mixing in the vessel,

$$F(t) = 1 - \epsilon^{-Q/V}$$

Thus, the F diagram shows how soon material entering the vessel begins to appear in the outflow and how long some portions of it may remain in the reaction zone. Where gravitational and surface-tension forces are negligible, geometrically similar systems at equal Reynolds numbers have identical F diagrams. Figure 15-1 shows typical F diagrams for various types of system.

The F diagram gives a complete picture of the distribution of residence times in a continuous reactor. For many purposes, the distribution function is represented sufficiently closely by a single parameter, the ratio of maximum to mean velocity or of mean to minimum residence time. This ratio has already been discussed in Chap. 3 as an index of flow pattern, and its reciprocal is plotted in Fig. 2-4. The ratio of mean to minimum time rather than the reverse is preferable in discussing residence times, and it will be called the *residence-time ratio* (RTR). For batch reactions and ideal plug flow, the value of the ratio is unity. For turbulent flow in pipes at "reduced velocities" (actual/critical) of more than about 4, it is approximately constant at

1.25. For streamline flow in cylindrical pipes and vessels, the value is 2.0. In a stirred vessel with theoretically perfect mixing, the value of the RTR would be infinity, since an infinitesimal portion of the inflow at any moment would instantly reach the outflow. For this reason, it is theoretically impossible for a continuous stirred-tank reactor to give a yield of product equal to that from a batch reactor, although in practice the difference may be so small as to be inappreciable.

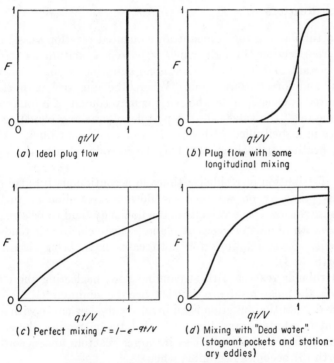

Fig. 15-1. F diagrams for reaction vessels.

Systems having identical F diagrams have also identical RTRs. The converse is not necessarily true, although where RTRs are identical in vessels of similar shape the F diagrams are not likely to be very different. It will be assumed that the agreement is sufficiently close for purposes of scaling up and that the RTR be taken as the practical criterion of residence-time distribution.

The rate of flow in a chemical reactor is often expressed in terms of space velocity, which is the volumetric rate of flow q divided by the reactor volume V. Its reciprocal is the mean residence time. In order to ensure that not only the main reaction but also all side reactions, known and unknown, will proceed to the same degree on the small and the large scale, it is not enough to have equal space velocities. There should also be equal RTRs and, ideally, identical F diagrams. The lower the RTR, the less the reaction volume necessary for a given throughput of reactants. Hence, it is desirable not only that the RTR shall be the same in the pilot unit and on the full scale but also that it shall be as low as is practicable.

Tubular Reactors

By a tubular reactor is meant an elongated reaction vessel of constant cross section through which is passed a continuous stream of fluid reactant and in which there is no agitation beyond that due to natural and forced convection. Where the quantity of heat to be transferred is considerable, the reactor may consist of a long, serpentine tube (e.g., the tube still of a petroleum-cracking plant) or a nest of tubes in a shell filled with a heating or cooling medium. Alternatively, heating or cooling coils may be introduced into the reaction space.

The distribution of residence times in a continuous tubular reactor depends primarily on whether the flow is streamline or turbulent. With gases at low space velocities, the flow may tend to be streamline, whereas with liquids or gases at higher space velocities it tends to be turbulent. Scale relations for both cases have been discussed by Bosworth (3, 4).

In a tubular reactor with streamline flow, neglecting the effect of diffusion, the mean residence time is twice the minimum (RTR = 2.0), and the distribution function is an inverse cube-law curve with a sharp cutoff at t_0. Lateral and longitudinal diffusion modify the shape of this curve, but the effect is less in larger reactors and according to Bosworth (3) becomes negligible when

$$d > 36 \sqrt{Dt_0} \qquad (15\text{-}1a)$$
$$l > 10d \qquad (15\text{-}1b)$$

where d = diameter of reactor
l = length of reactor
D = diffusion coefficient
t_0 = minimum residence time

all in consistent units. Above the minimum dimensions given by Eqs. (15-1a) and (15-1b), the distribution function and RTR are independent of reactor shape.

The condition for streamline flow may be written

$$d < 20 \sqrt{Sc \; Dt_0} \tag{15-2}$$

where Sc = Schmidt number.

For gases, where $Sc \approx 1.0$, Eq. (15-2) is incompatible with (15-1a), and the flow becomes turbulent before the minimum reactor size for negligible diffusion effect is reached. For liquids, Sc may be between 10^3 and 10^9, and the two equations are not incompatible.

Denbigh (11) has given a mathematical solution for the particular case of a tubular reactor carrying out a second-order reaction under conditions of streamline flow. It is based on Bosworth's analysis of the distribution function and does not allow for diffusion effects or side reactions of different orders.

In a tubular reactor with turbulent flow, the velocity distribution in the turbulent core is given by

$$v = v_{max} \left(\frac{r_{max} - r}{r_{max}} \right)^{1/m}$$

where v = fluid velocity at radius r

v_{max} = velocity at axis of reactor where $r = 0$

r_{max} = internal radius of reactor = $\frac{1}{2}d$

The following values of m and the corresponding RTRs are quoted by Bosworth from Ewald, Poschl, and Prandtl:

	m	RTR
Rough pipes..............................	5	1.32
Smooth pipes, Re $\geq$ 2,000 $\leq$ 100,000.........	7	1.22
Smooth pipes, Re > 100,000.................	8	1.19

Taking into account radial and longitudinal eddy diffusion, Bosworth derives a complicated equation for the distribution function (4). The shape of the distribution curve depends almost entirely on the ratio of length to diameter of the reactor and is affected only to a minor degree by the Reynolds number and roughness of the walls. Under no easily realizable conditions is the distribution of residence times independent of reactor shape where turbulent flow prevails.

The following conclusions may be drawn regarding the scaling up of homogenous chemical reactions carried out in continuous tubular reactors:

1. In order that the distribution function and RTR on the small and the large scale may be even approximately equal, it is necessary

that the dynamic regime in the two reactors shall be the same, either both streamline or both turbulent.

2. Turbulent flow in both reactors is preferable in that the RTR is lower and the distribution function less influenced by molecular diffusion.

3. At Reynolds numbers above the critical, the RTR may be taken as approximately constant at 1.25.

4. Under a turbulent regime, the distribution functions on the small and the large scale will be more nearly equal if the reactors are geometrically similar.

Subject to these qualifications, the corresponding throughput relations between model and prototype reactor are those giving equal mean residence times, and the appropriate scale equations are Eqs. (7-27a) to (7-27d).

In tubular reactors entirely filled with liquid or in reactors containing a vapor phase but in which the frictional pressure drop is negligible compared with the total absolute pressure of the system, the mean residence time under corresponding conditions is proportional to V/q_0, reactor volume divided by volumetric rate of input, even where there is a change of volume in the reactor and the true mean residence time is unknown. For vapor-phase reactions, this is no longer so when the length of the reactor is very great compared with its diameter and the pressure drop is substantial in relation to the absolute pressure. To obtain equal mean residence times in mixed or vapor-phase reactions, it is then necessary to lay down a further condition, that the pressure drop through the small-scale reactor shall equal that through the prototype. This in turn necessitates a departure from geometrical similarity, or at least a geometrical distortion such as was described in Chap. 3.

An example is the tube still as employed in the petroleum industry, where the absolute pressure at the inlet may be several times that at the outlet. V/q_0 is then no guide to the true mean residence time unless the initial and final pressures are also equal on the small and the large scale. The problem of designing a model tube still to reproduce the performance of a full-sized one, or vice versa, was discussed by Edgeworth Johnstone (18). If the fluid-friction factor is taken as a negative-power function of the Reynolds number,

$$f = K(\text{Re})^{-y}$$

then, from the D'Arcy formula for fluid friction in pipes with turbulent flow, the scale equations for homologous systems to give equal values of V/q_0 and equal pressure drops can be shown to be

$$1 = (d)^{(1+y)/(3-y)} \tag{15-3a}$$
$$q = (d)^{(7-y)/(3-y)} \tag{15-3b}$$

In fact, y is not entirely constant over the whole range of turbulent Reynolds numbers, but it is a fair approximation to take a mean value of 0.16 as proposed by Genereaux (105). Substituting this value, the above scale equation becomes

$$1 = d^{0.41} \tag{15-3c}$$
$$q = d^{2.41} \tag{15-3d}$$

In other words, the scale ratio normal to the direction of flow is greater than that parallel to the direction of flow, and the model tube still should be longer in proportion to its diameter than the prototype. This assumes a turbulent regime in both stills. Where there is stream-line flow in both, as might conceivably occur in vapor-phase cracking, it can be shown from Poiseuille's equation that the scale equations are

$$1 = d = q^{1/3} \tag{15-4}$$

i.e., there should be undistorted geometrical similarity with through-put proportional to reaction volume.

Example. A petroleum thermal cracking still contains 68 tubes 30 ft long by 3 in. ID followed by 78 tubes 30 ft long by 4 in. ID, arranged to carry two streams of oil in parallel. The throughput is 3,000 barrels/day of oil preheated to 300°F. It is desired to construct a model still having a throughput of 10 barrels/day and capable of reproducing the effects of variations in charging stock and operating conditions in the large still. Calculate the diameter and length of tube required in the model.

The large still has two streams in parallel, each having 34 three-inch and 39 four-inch tubes and cracking 1,500 barrels of charge per day. For each stream:

$$\text{Total length of 3-in. tube.......... 1,020 ft}$$
$$\text{Total length of 4-in. tube.......... 1,170 ft}$$

The model still will have a single stream; hence, $q = 1,500/10 = 150$.

1,500 barrels/day corresponds to over 2 fps in the 3-in. tube, which with pre-heated oil is well in the turbulent range. In the 4-in. section, turbulence is maintained by vaporization. Hence, Eqs. (15-3c) and (15-3d) may be used:

$$d = 150^{1/2.41} = 8.0$$
$$1 = 8.0^{0.41} = 2.35$$

Thus, the model still will contain:

$$\frac{1,020}{2.35} = 434 \text{ ft of } \tfrac{3}{8}\text{-in. ID tube}$$

$$\frac{1,170}{2.35} = 498 \text{ ft of } \tfrac{1}{2}\text{-in. ID tube}$$

To simulate the effect of the return bends on pressure drop, the model still should preferably also have 34 of the smaller and 39 of the larger tubes each $30/2.35 = 12.75$ ft long.

Checking the regime in the model, 10 barrels/day corresponds to about 0.95 fps at the entry to the $\frac{3}{8}$-in. tube, which for a viscosity of 1 centipoise is just in the turbulent region. The turbulence will increase as the oil heats up, and hence the assumption of a turbulent regime in both model and prototype is justified.

Catalytic Reactors

By a catalytic reactor is meant a continuous-flow tubular reactor containing a catalyst mass. The two main types of catalytic reactor are termed *fixed bed* and *fluidized* according as the catalyst mass is supported or freely suspended in the fluid.

In a catalytic reactor, it is possible to vary not only the particle size and shape of the catalyst but also its surface activity. Bosworth (2) has pointed out that by varying the catalyst activity in geometrically similar reactors it is theoretically possible to satisfy the requirements for both dynamic and chemical similarity. In Chap. 7, it was shown that where geometrical similarity extends to the catalyst grains the ratio of surface activities (α) necessary for simultaneous dynamic and chemical similarity is*

$$\alpha = \frac{1}{L}$$

The requirement is thus that the surface activity of the catalyst in the model reactor shall be L times that in the prototype, and it has already been pointed out that this relation has little practical utility since for economic reasons the large-scale reactor would normally be charged with a catalyst of the highest possible activity.

There is also a limit to the extent to which the activity per unit mass of a given catalyst can be increased by reducing the grain size. Thiele (25) found that for a solid catalyst of constant surface activity there is a critical grain size below which the activity per unit mass no longer increases. This is attributed to the diffusion of reactants throughout the catalyst mass where the particle size is very small. The activity per unit mass is determined by a dimensionless modulus which for first-order reactions may be written

$$\frac{X^2\alpha}{Dr}$$

* Equation (7-29d).

where X = linear dimension of grain (e.g., diameter of equivalent sphere)

α = surface activity of catalyst

D = diffusion coefficient of reaction through fluid

r = hydraulic mean radius of pores

For second-order reactions, the modulus is

$$\frac{X^2 \alpha a}{Dr}$$

where a = concentration of reactant in body of fluid.

In consistent units, the critical value of both moduli is about unity. Below this value, the catalyst activity per unit mass is constant; above it, the mass activity varies inversely as the modulus. An important consequence relates to the effect of temperature. An increase in temperature greatly increases the surface activity α and therefore the value of the modulus. If this is above unity, an increase corresponds to a reduction of catalyst activity per unit mass and hence the temperature coefficient and activation energy of the reaction will appear to be abnormally low. The same effect may occur if the modulus is below unity but a small (second-order) pore system exists in the surface of the main pores. A further consequence is that, where successive reactions occur, different values of the modulus will result in different relative proportions of the various products.

For the above reasons, the only reliable method of directly scaling up the performance of a catalytic reactor is for the small-scale reactor to constitute an element of the prototype, packed with catalyst identical in grain size and activity. That is, the pilot unit would consist of a tube containing the same depth of catalyst bed as the large-scale reactor and with external cooling, heating, or adiabatic jacket as required. Alternatively, a short experimental catalyst tube may be operated isothermally as a "differential reactor" and the results integrated mathematically by methods such as those of Hurt (17) or Dodd and Watson (12). The small-scale reactor is then a piece of laboratory apparatus rather than a model unit.

All of the above considerations apply equally to fixed-bed and fluidized reactors. Danckwerts, Jenkins, and Place (9) measured the distribution of gas residence times in a fluidized petroleum catalytic cracking regenerator 40 ft in diameter and obtained a distribution curve that was, in fact, similar to the distribution for streamline flow in a long pipe. This is not altogether surprising in view of the high kinematic viscosity of the hot gas and the small hydraulic mean radius of the channels between catalyst grains.

Continuous Stirred-tank Reactors

The distribution of residence times and the degree of completion of chemical reactions in continuous stirred-tank reactors have been treated mathematically by Macmullin and Weber (24), Kirillov (20), Brothman, Weber, and Barish (6), and Denbigh (10, 11). Eldridge and Piret (14) have extended and summarized the design equations for homogeneous chemical reaction of various orders, unidirectional, reversible, simultaneous, and consecutive. All are derived from a generic equation which may be written

$$A_n = A_{n-1} - \left(\frac{dA}{dt}\right)_n \theta_n \qquad (15\text{-}5)$$

where A = molar concentration of initial reactant
   t = time
   n = number of vessel from inlet end
   θ_n = mean residence time in nth vessel

As shown in Chap. 7, Eq. (7-2b), $-dA/dt$ depends upon the order of the reaction and the concentrations and activities of the reactants. Over-all integrations of Eq. (15-5) are given for zero-order and first-order unidirectional reactions; in other cases, a stepwise (vessel-to-vessel) solution is necessary.

To apply the design equation for continuous stirred-tank reactors, it is necessary to know the velocity constants of all the reactions taking place. In the case of consecutive reactions, the equations apply only when the distribution of residence times or F diagram is constant. Macmullin (23) has shown that, where consecutive reactions take place, batch, single-stage, and multistage continuous operation can lead to different relative proportions of reaction products.

The distribution of residence times in a series of continuous stirred-tank reactors depends far more on the number of stages and their capacity than on the flow pattern in each, since the time taken for complete mixing in each stage is short compared with the mean residence time. Therefore, provided that the Reynolds number is in all cases well in the turbulent range, volumetrically similar series of continuous stirred-tank reactors may be expected to give equal product compositions from a given feed when corresponding temperatures are equal and mean residence times in corresponding vessels are equal. From a chemical standpoint, there is no necessity for geometrical similarity, but it may be desirable to have geometrically similar vessels in order that heat-transfer coefficients and power consumption may be extrapolated directly, as described in Chaps. 12 and 14, respectively. The scale equations for homogeneous reactions are given in Chap. 7, Eqs. (7-27a) to (7-27d). These equations apply also to free-interface heterogeneous reactions (two liquid phases) which are subject to a chemical regime, i.e., provided the reaction rate is substantially independent of the rate of stirring over the range of Reynolds numbers between that in the model and that in the prototype.

For two-phase heterogeneous reactions subject to a dynamic regime (diffusion-controlled), the aim must be to secure approximately the same interfacial area per unit volume on the small and the large scale. Information on this subject is scanty, and in the light of present knowledge the best approximation is to operate the small- and large-scale reactors at equal power inputs per unit volume (see Chap. 14).

Transfer from Batch to Continuous Operation

It may sometimes be desirable to transfer a chemical reaction from batch operation in the laboratory or pilot plant to continuous operation on the large scale without the intervention of a continuous pilot plant. Jones (19) gives a convenient graphical method of designing a series of continuous stirred-tank reactors from a single batch-reaction curve, which, however, is strictly applicable only under the following conditions:

1. The product concentration in each element of reaction mixture is solely a function of time. (This is true only for zero- and first-order reactions or for reactions of higher orders that behave as first-order reactions owing to excess of other reactants.)

2. The curve of concentration vs. time has no maxima or minima. (This is not true where a reaction product is subject to subsequent decomposition.)

3. The reaction rates in batch and continuous systems are the same when the product concentrations are the same. (This is not true for simultaneous or consecutive reactions.)

4. Mixing in each reactor is instantaneous. (This is a reasonable approximation for normal mean residence times.)

Subject to these limitations, the graphical method permits a rapid computation in each reactor for any number of reactors, even where they are of different sizes [Denbigh (10) has pointed out that there may be advantages in having successive reactors of varying sizes, depending on the characteristics of the reaction]. As an approximate design method it is recommended by Jones for reactions of higher orders than first, when it is stated to give results that are slightly conservative. One advantage of the graphical method is that it is not necessary to know the velocity constant or even the order of the reaction. All that is required is a batch-reaction curve showing the concentration of the desired product or of some reactant as a function of time. The method can also be applied to any property such as density which is additive in mixtures, but for clarity it will be described in terms of product concentration.

The method of Jones depends upon the graphical solution of two simultaneous equations:
For the batch reaction:

$$\frac{dC}{dt} = \phi(C) \tag{15-6}$$

For the continuous reaction:

$$C_n = \phi'\left(\frac{dC}{dt}, C_{n-1}, \theta_n\right) \tag{15-7}$$

where C = concentration of product in reaction mixture
n = number of reactor
t = time
θ_n = mean residence time in nth reactor
ϕ, ϕ' = functions

The solution of Eq. (15-6) is obtained from the batch-reaction curve by plotting dC/dt (ordinate) against C (abscissa). This derivative curve is used for the graphical construction.

For a continuous stirred-tank reactor, Denbigh (10) has shown that for steady-state condition in the nth reactor

$$\frac{dC}{dt} = \frac{\Delta C}{\Delta t} = \frac{C_n - C_{n-1}}{\theta} \tag{15-8}$$

where ΔC and Δt represent over-all increments of concentration and time. The graphical solution of Eqs. (15-6) and (15-8) is effected as follows:

1. On the plot of dC/dt versus C, mark the point $dC/dt = 0$, $C = C_f$ = concentration of feed to the first reactor.

2. From this point, draw a straight line of slope $1/\theta_1$, where

θ_1 = mean residence time in first reactor
= liquid volume/feed rate (ordinate 1, abscissa = θ_1)

3. The intersection of this straight line with the batch curve of dC/dt versus C gives the concentration of the effluent from the first reactor = C_1.

4. From the point $dC/dt = 0$, $C = C_1$, draw a straight line of slope $1/\theta_2$. Its intersection with the batch curve gives the concentration in the effluent from the second reactor $= C_2$.

5. Repeat the process for the desired number of reactors to obtain the concentration of the final effluent.

The graphical calculation is illustrated in Fig. 15-2, which is copied from Jones's paper.

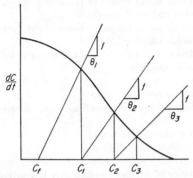

Fig. 15-2. Jones's method of computing continuous reactors in series from the batch-reaction curve. [*R. W. Jones, Chem. Eng. Progr.*, **47**: 46 (1951).]

Accurate design calculations for reactions of second or higher orders, or for simultaneous or consecutive reactions, require the stepwise application of the equations of Macmullen and Weber (24) or Eldridge and Piret (14). The problem then reduces to the mathematical manipulation of laboratory data and specified values of the boundary conditions rather than the scaling up of pilot-plant results. If the necessary data are available, it may even be economically advantageous to by-pass the pilot stage altogether.

SYMBOLS IN CHAPTER 15

A = molar concentration of initial reactant in main body of fluid
C = concentration of reaction product
ΔC = over-all increment of concentration in one reaction vessel
D = diffusion coefficient
d = diameter of reactor
E = activation energy
F = distribution function
ΔF = free energy at standard activity
f = fluid-friction factor
K = equilibrium const of chemical reaction
K = a const
k = velocity const of chemical reaction

L = linear dimension

l = length of reactor

m = an exponent

n = number of reaction vessel, counting from the inlet end

Q = total volume of outflow up to time t

q = volumetric rate of flow

q_0 = volumetric rate of input (to first vessel)

R = gas const

r = radial distance from axis of reactor

r = hydraulic mean radius of pores in catalyst mass

r_{max} = internal radius of reactor = $\frac{1}{2}d$

r = hydraulic mean radius of pores in catalyst mass

Re = Reynolds number

RTR = residence-time ratio = ratio of mean to minimum residence time

Sc = Schmidt number

t = time

t_0 = minimum residence time

Δt = residence time corresponding to an increment of concentration ΔC

V = volume of reaction vessel

v = fluid velocity at radius r

v_{max} = fluid velocity at axis of reactor where $r = 0$

X = grain size of catalyst = diameter of equivalent sphere

y = an exponent in Eqs. (15-3a) and (15-3b)

α = surface activity of catalyst

ϵ = 2.718 $\cdots$

θ_n = mean residence time in nth vessel

ϕ, ϕ' = functions

Subscripts

$n, n + 1$, etc. = pertaining to the nth, $(n + 1)$st reactor, etc.

0 = at inlet to (first) reactor, *except* in t_0.

CHAPTER 16

FURNACES AND KILNS

The design and operation of furnaces and kilns present many complex problems that are incapable of exact or even approximate mathematical solution. Consequently, models are widely used in this field. They are employed both as aids to the design of new furnaces and in the study of existing installations to improve their efficiency. The principal aspects of furnace performance in which model studies can be of value are:
1. Furnace aerodynamics
2. Geometry of flames
3. Behavior of solid-fuel beds
4. Heat transfer
5. Physical and chemical changes in the charge
6. Flow of molten materials and slags
7. Flow of cooling water
8. Behavior of refractories

Three general types of furnace model have been used in studying these phenomena:

Cold models, in which the flow of combustion gases, and sometimes that of molten products, is simulated by cold fluids such as air or water.

Warm or medium-temperature models, in which heat-transfer and phase changes are simulated at relatively low temperatures in non-homologous systems. For example, the fusion of pig iron and steel scrap has been simulated by the melting of model pigs made from paraffin wax (65).

Hot or high-temperature models, which operate at temperatures of the same order as those in the prototype furnace. They may be homologous with the latter, in which case they process the same materials and constitute in some degree working models of full-sized furnaces. For example, both blast furnaces and open-hearth furnaces have been studied by means of scale models which actually produce pig iron or steel, respectively (28, 35 to 37).   Alternatively, the charge may be replaced by a measured heat sink with controllable surface temperature.

The experimental apparatus is often not a complete model of the prototype furnace, but rather a model element as defined in Chap. 3. The so-called "slice model" of a reverberatory or shaft furnace is a model element geometrically derived from a complete model by slicing it parallel to the direction of flow.   Owing to the large size of many industrial furnaces, full-scale elements are rarely met with, but as described later, it is sometimes advantageous to use the furnace itself as a full-scale cold model.

Furnace Aerodynamics

Most of the important features of gas flow in furnaces can be simulated on the small scale without the application of heat.   The use of cold models for this purpose has been greatly developed in recent years; for example, steel companies now use them regularly in the exploration of open-hearth furnace design.

The pioneers in experimenting with cold furnace models were Groume-Grjimailo (40) and Rosin (59).   Groume-Grjimailo used oil and water to simulate buoyancy effects with hot and cold gases, and his findings were mainly qualitative.   The most important conclusion was that in order to provide a uniform distribution of heat in a furnace the combustion gases should travel downward as they cool.   Buoyancy forces will then automatically prevent the formation of cold pockets or the by-passing of hot gases.   Conversely, where the gases are being heated (as when cold air is drawn through a regenerator box), buoyancy forces will tend to bring about a uniform distribution when the flow is upward.   The experiments of Groume-Grjimailo were conducted at fluid velocities far below those required for dynamic similarity, and consequently they failed entirely to simulate the effects of turbulence.   Rosin (59) developed the water model for the study of flow distribution in complex combustion systems and maintained dynamic similarity in order to ensure equivalent turbulence in the model.

More recent workers who have used cold-water models include Chesters and his collaborators in investigations of the open-hearth furnace (30), Newby in studying the side-blown Bessemer converter (53), and, in an allied field, Watson and Clarke investigating the combustion chambers of gas turbines (71).

A cold model is capable of giving both qualitative and quantitative information about flow conditions in the prototype furnace.   Quantitative data can be obtained on pressure drops and velocity profiles in different parts of the furnace. Qualitative indications are obtainable of turbulence patterns in the furnace gases and regions in which flame impingement or excessive erosion of refractories may occur.

Cold models are generally operated with either air or water as the working fluid. The kinematic viscosity of hot furnace gases is of the order of 12 times that of cold air and 130 times that of water. Hence, for dynamic similarity in a one-twelfth scale model ($L = 12$), the cold-air velocity should be equal to that in the prototype, whereas with water as the fluid the velocity is required to be less than one-tenth as great. Owing to their lower corresponding velocities, water models are more suitable for the visual observation of flow patterns, e.g., by the injection of a pulse of dye. Air models are easier to construct and more suitable for impingement studies and the determination of velocity profiles by a pitot-tube traverse. Most furnace laboratories employ both types.

Pressure drops in various parts of a model-furnace system are determined and scaled up by methods similar to those described in Chap. 10, *Ducts and Flow Passages*. Dynamic similarity requires that the Reynolds numbers in a model and prototype shall be equal, and with an air model this is not always practicable. However, as was pointed out in Chap. 10, provided that the Reynolds number in the model is not less than three or four times the critical value, the change in flow pattern and D'Arcy friction factor for any further increase in the Reynolds number is comparatively slight. The model results can therefore be extrapolated to higher Reynolds numbers, assuming a constant friction factor with relatively little error. That is, above a "reduced velocity" of 3 or 4, the pressure coefficient for geometrically similar systems may be taken as constant, and the scale equation becomes

$$\frac{\Delta p}{\rho v^2} = 1$$

This assumption is likely to give conservative results; i.e., the predicted pressure drop in the prototype may be slightly higher than the actual pressure drop.

Various devices are employed to render visible the flow pattern in a furnace. For visual observation, a scale model or a model element representing a longitudinal slice of the furnace is constructed in transparent material, e.g., Lucite or Perspex. Where water is the working fluid, periodic pulses of dye may be injected into the stream, as already mentioned. Alternatively, the water may have suspended in it a small proportion of aluminum "bronze powder" (minute flakes of aluminum as used in paints). Fine air bubbles dispersed in the water can also act as flow indicators. Such techniques have been used, for example, by Chester and coworkers in the study of open-hearth furnaces (30) and by Watson and Clarke in relation to gas-turbine combustion chambers (71). The model is illuminated by a thin, parallel beam of light so as to reveal a cross section of the flow.

In cold-air models, the flow may be rendered visible by the addition to the air stream of a light powder such as balsa wood dust. A method that has been used in the study of impingement wear on the roof of an open-hearth furnace is the "sticky-dust" technique (48). The inside of the roof of the model is coated with a sticky material, and a standard quantity of dust is injected into the air stream representing the fuel gas. At dynamically similar flow rates, the distribution of this dust over the roof has been shown to correlate closely with the wear in an actual furnace, the best design being that which distributes the dust most uniformly.

The velocity distribution and flow pattern in an air model may be explored directly by means of pitot tubes. One technique that has been fruitfully employed is to use the furnace itself as a cold model on the full scale. By sealing up all joints and cracks with brown paper or water glass and using an induced-draft fan to draw through the furnace enough cold air to give a Reynolds number equal to that in the hot system, a very exact reproduction of the working flow pattern is obtained. All roughness and irregularities are exactly reproduced in the "model," and there is the further advantage that investigators can enter the furnace and explore the flow pattern by means of sensitive air-velocity meters. Changes in the geometry can be introduced by placing wooden shapes at suitable points. This technique has been applied by Gooding and Thring to a glass tank (39) and by Newby, Collins, and Leys to an open-hearth furnace (54). The chief practical drawback is the time limit normally imposed by the fact that the furnace is wanted for production.

In all the cold models discussed so far, it has been assumed that buoyancy effects are negligible, which is true of systems in which the gas velocity is high or the temperature differences are not very great. It is not true, however, of systems such as regenerators in which air is heated from atmospheric temperature to over 1000°C. Groume-Grjimailo's technique of using two immiscible liquids of different densities is not suitable for high-velocity turbulent systems. Rosin (59) has used salt briquettes to represent solid fuel in an inverted water model of a solid-fuel furnace, the density rise as the salt dissolved simulating the reduction of gas density in the prototype furnace due to rise of temperature of the combustion products.   This technique has a limited application; the density change from 1.0 to 1.4 in the model is much less than that occurring in the prototype, where the hot gases may have only one-quarter the density of the cold air, and the method cannot be used when the quantity of water flowing in a model is such that recycling is necessary.   Horn and Thring (41) are developing a technique involving the use of a suspension of finely ground magnetite, which can give a density change of more than 2:1 with water and which can be filtered off for recycling.

Two-phase Models.   In the cold models described so far, the kinetic system has consisted of a single fluid phase in motion.   Many of the actual processes occurring in furnaces are two-phase, the second phase being either solid as in the combustion of fuel or liquid as in the refining of molten metal.   As already mentioned, Rosin (59) used briquettes of salt to represent solid fuel in an inverted water furnace. The increased density of the salt solution represented the decreasing density of the gases passing through the fuel bed and the dissolution of the briquettes represented the combustion of the fuel.   The latter process is analogous to combustion only when the fuel temperatures in the prototype are so high that the reaction rate is subject to a dynamic regime, i.e., controlled by rate of flow of the combustion air.

Newby (53) has made a study of the gas flow in a side-blown converter resulting from the impingement of high-velocity air jets on the surface of the molten steel.   Similarity theory in its simplest form requires that both the Froude number and the ratio of gas to liquid densities should be the same in model and prototype.   A more careful analysis indicates that the flow pattern must depend on a single dimensionless group, or *modified Froude number*, consisting of the product of the Froude number and the ratio of gas to liquid density,

$$\frac{\rho_g v^2}{\rho_l L g} = \text{const}$$

where ρ_g, ρ_l = density of gas, liquid

v = gas velocity

L = linear dimension

g = acceleration of gravity

This modified Froude number represents the ratio of the inertial force in the gas to the gravitational force acting on the liquid.   It is thus possible to use any liquid and any gas in the model provided the gas velocities in model and prototype are such as to satisfy the above relation.   In two models, one containing water and one mercury, in which the modified Froude numbers were equal, the disturbance of the liquid surface in both systems was similar, whereas at equal gas Froude numbers the patterns were quite different. Newby finally modeled the disturbed liquid surface in Plasticine and operated the model with Reynolds similarity using water to represent the combustion gases, in order to study the flow pattern visually.

Water has been used to simulate the flow of steel running out of an open-hearth furnace into a ladle and finally splashing on the bottom of the mold, another two-phase system. Here the similarity criterion for geometrically similar arrangements is equality of Froude numbers (unmodified). The flow of steel from a furnace is sometimes divided by a forked channel to feed two ladles, a refractory brick being inserted in the flow to regulate the quantity to each ladle. The water model can be used to find the best shape and location for the brick.

Geometry of Flames

Model studies have been used to predict the size and shape of flames in furnaces and the degree of impingement on the charge or on the roof and walls of the furnace. Combustion itself is a chemical process, but the fuel and air are brought together by the physical processes of turbulence and diffusion. Hence, before model studies are undertaken, it is necessary to consider the question of regime as discussed in Chap. 6, i.e., whether the rate-controlling factors are chemical, aerodynamic, or both.

The structure and properties of furnace flames have been discussed in detail by Thring (67), Gaydon and Wolfhard (38), and others. Flames are produced by the combustion of gaseous or vaporized fuels, atomized liquid fuels, and pulverized solid fuels suspended in a stream of air. The last two behave similarly to gaseous fuels so far as the shape and structure of the flame are concerned. There are two principal types of flame, *premixed flames* and *diffusion flames*. In a premixed flame, the gaseous or atomized fuel is mixed with the necessary combustion air before issuing from the burner. Diffusion flames result when the fuel and air are introduced separately into the combustion space, and the fuel burns as it mixes with the air.

In premixed flames the physical mixing of fuel and air has taken place outside the combustion space. When the flame is streamline or laminar, the rate of burning depends only upon the chemical-reaction velocity, and the shape of the flame front is governed by the ratio of the burning velocity to the gas-flow velocity in the tube. When the gas-flow velocity becomes so high that the flame is turbulent, however, the burning velocity is found to increase to an extent depending upon the Reynolds number. The combustion process in the streamline premixed flame is therefore subject to a chemical regime, while in the turbulent premixed flame there is a mixed regime.

In diffusion flames the rate of burning depends upon the rate at which oxygen reaches the fuel, and this in turn depends upon the flow pattern and degree of turbulence in the combustion space. A diffusion flame, whether streamline or turbulent, is therefore subject to a dynamic regime.

Pure premixed flames in which the fuel is mixed with sufficient air for complete combustion are unstable, being liable to strike back or blow off the burner with small changes in gas velocity. They are rarely used in industrial furnaces. In the ordinary bunsen burner, the fuel is premixed with only a part of the air needed for combustion. The bunsen flame is thus a compound flame; the inner blue cone is a streamline premixed flame in which some of the fuel reacts with the premixed or primary air, while the outer cone is a diffusion flame in which the remainder of the fuel reacts with the surrounding atmosphere. The bunsen flame is a good example of a mixed chemical-dynamic regime which cannot be geometrically scaled up even though the regime remains the same (either streamline or turbulent) throughout.

In a premixed flame such as the inner cone of a bunsen flame, the flame front is relatively thin owing to the high rate of combustion. The angle of the cone at any point is such that the gas-velocity component normal to its surface is equal to the rate of flame propagation in the mixture (measurement of the cone angle is one method of determining the rate of flame propagation) (38). Hence, for a given temperature, pressure, and mixture composition and a given shape of burner, all streamline premixed flames will form geometrically similar cones when the gas velocities are equal. The flame fronts will, however, be of constant thickness and not proportional to the flame diameter.

Limits to the possible size of a bunsen burner are set by the *quenching distance*. For reasons that are not fully understood, a flame will not burn within a certain small distance of the burner wall, the value depending on the temperature, pressure, and composition of the mixture. It is not solely a cooling effect, since the distance is little affected by the thermal conductivity of the burner tube. In the ordinary laboratory bunsen burner, this quenching effect prevents the flame from striking back along the walls of the burner where the gas velocity is low. As the diameter of the burner is increased, however, the boundary region of low gas velocity comes to extend beyond the quenching distance and the flame will strike back down the burner tube unless the gas velocity is also increased. The maximum possible gas velocity, and hence the maximum burner diameter, is fixed by the gas-supply pressure. For city gas at normal pressures, the maximum diameter of a simple bunsen burner is about ¾ in.

Large bunsen flames may be stabilized by means of a gauze or grid as used in the Meker burner. Each element of the grid is in effect a miniature bunsen burner with its own small inner cone. The effect of the grid is to render the total height of the premixed flame region small compared with the height of the subsequent diffusion flame, so that within limits Meker flames may be treated as pure diffusion flames.

Both premixed and diffusion flames may be either *streamline* or *turbulent*. Streamline flames are silent and in still air have a fairly sharp outline. Turbulent flames are noisy, and their outlined is blurred. Where the Reynolds number in the burner tube is less than 2,300, the flame is streamline; above 3,200, it is fully turbulent; at intermediate values, the flame alternates between the streamline and turbulent forms (50). In order to have either fully streamline or fully turbulent conditions, the burner tube must have a minimum length L_m which is related to the diameter d and the Reynolds number (38),

$$\frac{L_m}{d} = 0.03 \text{ Re}$$

Such long burner tubes are seldom found except in research work. Hence, even at low gas velocities, the flame from an ordinary burner is never fully streamline. It may nevertheless be regarded as having a flame shape corresponding to streamline conditions, so that the regime is chemical.

Premixed or predominantly premixed flames tend to be short, hot, and only slightly luminous. Their heat is transferred to the charge mainly by convection, and since convection coefficients in gases are low, this limits the maximum rate of heating that can be achieved. Much higher heat-transfer rates are attainable by radiation, and for this reason the luminous diffusion flame is preferred for many industrial operations. A further advantage of the diffusion flame is that the sensible heat of the combustion gases leaving the furnace can be largely recovered by using it to preheat the combustion air and the fuel, if gaseous, in separate regenerators or recuperators. It would not be safe to heat the premixed fuel gas and air in this manner, owing to the risk of explosion.

In diffusion flames, there are two "resistances" (in the sense defined in Chap. 6) which tend to reduce the rate of combustion: a conversion, or chemical, resistance associated with the activation energy of the reaction, and a diffusion, or dynamic, resistance dependent upon the aerodynamics of the system. For reasons of fuel economy, industrial diffusion flames employ preheated combustion air and fuel wherever possible, and because of the high temperature coefficient of chemical-reaction velocities the chemical resistance in such flames is low. At high temperatures, also, the viscosity of gases is increased, and the degree of turbulence is consequently less than when they are cold. Hence, the controlling factor is the time taken for fuel and combustion air to be brought together by eddy and molecular diffusion. Diffusion flames at normal pressures are thus subject to a dynamic regime, and it can be assumed that combustion is complete as soon as fuel and air are intimately and completely mixed in stoichiometric proportions. This assumption is generally a good approximation even when the fuel and air are not preheated unless the degree of turbulence in the combustion space is unusually high.

In so far as diffusion flames are subject to a dynamic regime, they will tend to be geometrically similar at equal values of the Reynolds number for both fuel and air streams. For homologous systems, the scale equation is

$$\mathbf{v} = \frac{1}{\mathbf{L}} \tag{16-1}$$

This relation can be applied over only a limited range of scale ratios because the air and gas velocities in a small-scale model become impracticably high. Further, at high velocities, it is possible that the turbulence may be so great that chemical resistance becomes appreciable and the regime is no longer purely dynamic.

The experimental criterion of a chemical regime is that the flame length is independent of the degree of turbulence but varies markedly with the temperature of the fuel and air. The criterion of a dynamic regime is that the flame length is only slightly dependent on the fuel and air temperatures. It may or may not vary with the gas velocity and degree of turbulence according to the Reynolds number. If, for example, a diffusion flame from a given orifice has a streamline flow pattern, its length is nearly proportional to the gas velocity. At high Reynolds numbers, where the flow is already fully turbulent, a further increase in gas velocity will not affect the flame length. Where the flame length is affected by both preheat and turbulence, a mixed regime is indicated. This may occur when:

1. A premixed flame is fully turbulent,
2. A streamline premixed flame has insufficient primary air for complete combustion and must acquire the balance by diffusion, or

3. A diffusion flame is highly turbulent, with fuel and air supplied at a low temperature.

In neither case can any predictions regarding flame geometry be made from the behavior of a model burner.

Baron (26) has made a theoretical study of the free turbulent diffusion flame employing a gaseous fuel. Assuming instantaneous combustion, the adiabatic flame temperature, and negligible buoyancy effects, he derived equations for the length and shape of the flame which are stated to agree with practice. For flame length or height,

$$\frac{l}{d} = \frac{5.2}{y} \sqrt{\frac{T_f}{\alpha T_0}} \left[y + (1 - y) \right] \frac{M_s}{M_0} \tag{16-2}$$

where l = flame length
d = nozzle diameter
y = mole fraction of fuel gas in stoichiometric combustion mixture
α = ratio of moles in stoichiometric mixture after combustion to moles before combustion
T_0 = absolute temperature of fuel gas to nozzle
T_f = calculated adiabatic flame temperature
M_s = mole weight of combustion air = 28.8 (unless enriched with oxygen)
M_0 = mole weight of fuel gas

The shape of the flame is given by the equation

$$\frac{D}{x} = 0.29 \sqrt{ln \frac{l}{x}} \tag{16-3}$$

where D = diameter of flame and x = distance from nozzle.

Equation (16-3) defines a smooth envelope corresponding to the mean shape of the flickering turbulent flame. Equation (16-2) contains no term for the gas velocity and therefore applies only to fully turbulent systems, where the flame length is independent of gas velocity. In homologous systems, the flame length is then proportional to the nozzle diameter only (l/d constant), so that all such flames should be geometrically similar, provided that instantaneous combustion can be assumed. With very rapid mixing of fuel and air due to high turbulence, it is possible that the rate of reaction or chemical resistance might have an appreciable influence on the total rate, especially where the fuel and air are not preheated. Under these conditions, fully turbulent diffusion flames would be geometrically similar only at equal gas velocities. Since it is not always certain whether the chemical resistance is appreciable in any given case, it is safer to accept the above limitation for all turbulent diffusion flames. This limits the scale-up ratio, since for a given gas velocity there is a minimum size of model burner that will give full turbulence (three to four times the critical Reynolds number).

So far the discussion has been limited to free or nonenclosed flames, and from the review of factors affecting flame geometry the following conclusions may be drawn. They presuppose homologous systems, i.e., identical fuel compositions, fuel/air ratios, and initial temperatures, and geometrically similar burner tubes:

1. Within the limits imposed by flashing back and quenching, streamline premixed flames are geometrically similar at equal gas velocities.

2. Streamline and partly turbulent diffusion flames are geometrically similar at equal Reynolds numbers. The scale-up relation is given by Eq. (16-3).

3. Fully turbulent diffusion flames are geometrically similar at equal gas velocities and approximately similar irrespective of gas velocity unless the reaction rate is unusually low.

4. Partially turbulent premixed flames are subject to a mixed regime. The shape of the cone depends upon both the absolute gas velocity and the velocity profile. The latter is a function of the Reynolds number, so that it is impossible to have both equal gas velocities and similar velocity profiles in burners of different sizes. Hence, a strict geometrical scale-up is not possible.

5. Fully turbulent premixed flames and compound premixed diffusion flames such as the bunsen flame are subject to a mixed regime and cannot be geometrically scaled up unless by means of a grid the inner premixed flame is reduced to negligible dimensions compared with the outer diffusion flame. The latter may then be scaled up as a pure diffusion provided the flow regime does not change.

Flames in furnaces and kilns are classed as enclosed flames; i.e., their geometry is modified by the walls of the enclosure. Further, whereas a free flame can entrain an unlimited amount of air, the combustion air supplied to a furnace is generally restricted in order to give high-temperature combustion gases. The jet of fuel gas may then, owing to its momentum, entrain a larger mass of stationary gas than corresponds to the air supply. This leads to the phenomenon of recirculation in the combustion chamber. The effect is illustrated diagrammatically in Fig. 16-1. Before the jet of hot gases reaches the furnace wall, it has entrained all the air and a quantity of combustion gases that have lost their momentum through contact with the wall.

If it can be assumed that a dynamic regime prevails and combustion is almost instantaneous once the air and fuel are mixed, then under conditions of recirculation the flame will not impinge upon the furnace wall but will assume a shape something like that shown in the figure. It is sometimes useful to be able to predict the shape and length of the flame by means of model experiments.

In the following discussion, it is assumed that the fuel is a gas. In practice, a suspension of atomized liquid or solid fuel in air or steam may be treated as a gas.

Consider a furnace heated by a jet of gaseous fuel which issues from a burner nozzle and meets a separate stream of air inside the combustion chamber, the burnt gases being withdrawn to a flue. This is an

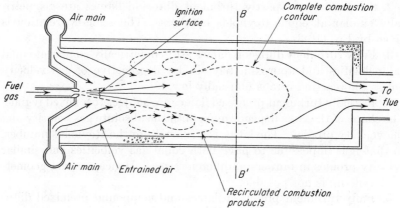

Fig. 16-1. Internal circulation and flame contour in a furnace.

example of the enclosed diffusion flame as commonly employed in industrial furnaces. Let

N = actual mass ratio of (air + fuel)/fuel

N_0 = stoichiometric mass ratio of (air + fuel)/fuel

[percentage excess air = $100(N - N_0)/(N_0 - 1)$].

v_f = mean velocity of burnt gases in region where temperature has become uniform and equal to T_f

ρ_f, μ_f = density and viscosity of burnt gases at temperature T_f

L = linear dimension along combustion chamber

N and N_0 are ready-made dimensionless similarity criteria. The first governs the relative mass-flow rates in the burner nozzle and combustion chamber; the second fixes the percentage excess air and hence the mixing length for a given value of N (instead of N_0, the excess air ratio could be taken as one criterion). The flow pattern in the later stages of combustion and afterward is fixed by the Reynolds number, and hence $\rho_f v_f L/\mu_f$ constitutes a third criterion.

The flow pattern in a furnace is complicated by the large rise in temperature of the gases in the first part of the combustion chamber near the burner nozzle. In this region, the jet of fuel gas is expanding as it entrains combustion air, mixes with it, and burns. The temperature across the flow path is far from uniform, and the concept of a mean velocity has no application, since the flow consists of a high-velocity jet surrounded by relatively stagnant air. Thus, equality of Reynolds number in the later stages of combustion is no criterion of similarity in the early stages, and it is conditions in the early stages that largely determine the shape and size of the flame.

Thring and Newby (68) derived a similarity criterion for the early part of the flame by assuming conservation of momentum in the free-jet region, before the expanding jet impinges on the combustion-chamber walls. Consider a free jet of fluid of density ρ_0 which entrains a surrounding fluid of density ρ_s. Except very close to the nozzle, the velocity profile across a free jet is constant, so that one can write

$$\rho_{rx} = \rho_x \phi_1(\eta)$$
$$u_{rx} = u_x \phi_2(\eta)$$

where x = distance along axis of jet from an imaginary point source at apex

r = radial distance of a point P from axis of jet at distance x

ρ_{rx} = density of mixture at P

ρ_x = density of mixture on axis of jet at distance x

u_{rx} = local velocity of mixture at P

u_x = local velocity of mixture on axis of jet at x

$\eta = r/x$

Let G = jet momentum, defined as mass rate of flow of fluid times velocity (jet momentum has the dimensions of a force)

It is assumed that the surrounding fluid enters at a negligible velocity so that all the momentum of the expanding jet is derived from the momentum of the nozzle fluid, and that this remains constant over the free-jet region.

The jet momentum across an annular element of radius r and width dr is given by

$$G_r = 2u_{rx}^2 \rho_{rx} \pi r \, dr$$

If r_e is the radius to the edge of the jet, the total jet momentum across the plane x is

$$G = 2 \int_0^{r_e} u_{rx}{}^2 \rho_{rx} \pi r \, dr$$

$$= 2u_x{}^2 \rho_x \pi x^2 \int_0^{\eta_e} \phi_1(\eta) \phi_2{}^2(\eta) \eta \, d\eta$$

$$= u_x{}^2 \rho_x \pi x^2 K_{12}$$

where $K_{12} = 2 \int_0^{\eta_e} \phi_1(\eta) \phi_2{}^2(\eta) \eta \, d\eta$, which is independent of x. Hence,

$$u_x = \sqrt{\frac{G}{\rho_x x^2}} \, \frac{1}{\sqrt{\pi K_{12}}} \tag{16-4}$$

Similarly, the total mass flow of the jet is given by

$$M = 2 \int_0^{r_e} u_{rx} \rho_{rx} \pi r \, dr$$

$$= 2u_x \rho_x \pi x^2 \int_0^{\eta_e} \phi_1(\eta) \phi_2(\eta) \eta \, d\eta$$

$$= u_x \rho_x \pi x^2 K_{11}$$

where $K_{11} = 2 \int_0^{\eta_e} \phi_1(\eta) \phi_2(\eta) \eta \, d\eta$, also independent of x. Substituting for u_x from Eq. (16-4),

$$M = x \sqrt{G \rho_x} \sqrt{\frac{\pi K_{11}{}^2}{K_{12}}}$$

or

$$\frac{M}{x} = \sqrt{G \rho_x} \, K_3$$

The original jet of density ρ_0 soon entrains many times its own mass of the surrounding air and rapidly heats up. Hence, the density ρ_x at the axis of the jet soon approximates to the density ρ_f of the final combustion gases at flame temperature. Thus, to a reasonable approximation one may write

$$\frac{M}{x} = \sqrt{G \rho_f} \, K_3 \tag{16-5}$$

where K_3 is a constant depending only on the jet angle, which in turn depends upon the nozzle geometry.

From (16-5), the total mass flow of original nozzle fluid plus entrained fluid at distance L from the origin is

$$M = K_3 \sqrt{G \rho_f} \, L$$

If $\quad a$ = cross-sectional area of the jet nozzle

$\quad\quad M_0$ = mass-flow rate of nozzle fluid

the mean velocity at the nozzle is $M_0/\rho_0 a$, and since G is constant,

$$M_0 = \sqrt{G\rho_0 a}$$

Hence, the ratio of total jet fluid to nozzle fluid at length of travel L is

$$\frac{M}{M_0} = K_3 \frac{L}{\sqrt{a}} \sqrt{\frac{\rho_f}{\rho_0}} \tag{16-6}$$

K_3 being a geometrical constant.

Equation (16-6) gives the fourth dimensionless criterion affecting the geometry of the flame. Putting l/L as a shape factor, the complete generalized dimensionless equation for the enclosed diffusion flame is

$$\frac{l}{L} = \phi\left(N, N_0, \frac{\rho_f v_f L}{\mu_f}, \frac{L^2 \rho_f}{a\rho_0}\right) \tag{16-7}$$

In most industrial furnaces, the Reynolds number is of the order of 100,000, and in this range a considerable variation has little influence on the flow pattern. For example, a model furnace may be operated at Reynolds numbers as low as one-tenth those in the prototype without sensibly impairing the similarity of flame contours. Hence, in practice the Reynolds criterion may be neglected, and Eq. (16-7) becomes

$$\frac{l}{L} = \phi\left(N, N_0, \frac{L^2 \rho_f}{a\rho_0}\right) \tag{16-8}$$

Assuming constant values of N and N_0, this gives the scale equation

$$\mathbf{a} = \frac{\mathbf{L}^2 \varrho_f}{\varrho_0} \tag{16-8a}$$

(The scale relation is given in terms of nozzle cross-sectional areas rather than diameters because burner nozzles are not necessarily cylindrical.)

Equation (16-8a) shows that two geometrically similar furnaces in which the fuel and flame temperatures are different will not give similar flame contours unless the scale of the burner nozzles differs from that of the combustion chambers. Where the flame contour in a prototype furnace is to be predicted by means of a cold-air model, then the scale of the model burner nozzle must be changed still more. If ρ_a be the density of the air used as both nozzle fluid and surrounding fluid in the model and ρ_0', ρ_f' be the densities of nozzle fluid and combustion gases in the prototype, then $\varrho_f = \rho_f'/\rho_a$ and $\varrho_0 = \rho_0'/\rho_a$. Equation (16-8a) thus becomes

$$\mathbf{a} = \mathbf{L}^2 \frac{\rho_f'}{\rho_0'} \tag{16-8b}$$

The nozzle diameter is a larger fraction of the combustion-chamber diameter in the model than in the prototype; i.e., the model is geometrically distorted.

Thring and Newby (68) also considered the case where atomized fuel oil enters a furnace carried by a jet of steam or compressed air traveling at ultrasonic velocity. They showed that in scaling down to a cold-air model the cross section of the full-sized burner nozzle should be taken, not as the actual cross section of the ultrasonic jet nozzle, but as the cross section of the carrier jet when it has expanded to atmospheric pressure without loss of momentum flux. The momentum flux, and hence the jet velocity and equivalent area, can be calculated either from the total heat of steam and an assumed nozzle efficiency or from direct measurements of the backward thrust on the burner. By this means, and using a cold model to a linear scale of 10, Thring and Newby were able to predict the combustion-gas compositions along the length of an oil-fired furnace with a standard deviation of about 15 per cent. In the model, methane was added to the nozzle fluid as a tracer gas and estimated by infrared absorption, as already mentioned.

Example. A furnace of internal cross section 12 ft square is fired by low-pressure fuel gas issuing from a nozzle 4 in. in diameter. The fuel gas is fired cold and has a density of 0.040 lb/ft.[3] The mean density of the hot combustion gases in the furnace is 0.008 lb/ft.[3] It is desired to predict the flame contour by means of a cold-air model 18 in. square. What should the diameter of the nozzle be in the model?

$$L = \frac{12}{1.5} = 8$$

$$\rho_0' = 0.040 \qquad \rho_f' = 0.008$$

From Eq. (16-8b),

$$a = 64 \frac{0.008}{0.040} = 12.8$$

The ratio of diameters $d = \sqrt{a} = 3.57$. Hence, the diameter of the model nozzle should be

$$\frac{4}{3.57} = 1.12 \text{ in.}$$

Behavior of Solid Fuel Beds

Three aspects of the behavior of solid fuel beds have been investigated by means of models: the pressure drop through the beds, the segregation of the solid particles when charged to the furnace, and their movement as combustion takes place.

Rosin (59) used water models to evaluate the pressure drop through beds of solid fuel. The similarity criterion is equality of Reynolds number, the characteristic linear dimension being taken either as the mean particle diameter or, better, as the hydraulic diameter of the voids through which the combustion gases must pass. The hydraulic diameter is equal to the volume of voids per cubic foot of fuel bed divided by the solid surface per cubic foot, termed the *specific surface*. It is seldom possible to provide exactly the same particle-size distribution or particle shape in the model as in the prototype furnace, and where these are different, the hydraulic diameter is a more correct measure of linear scale than the mean particle diameter.

At equal Reynolds number in model and prototype, the pressure coefficients are equal, and the ratio of pressure drops is given by Eq. (7-11g). The experimental equipment for the test is similar to that illustrated in Fig. 10-2.

The segregation of solid particles by density and size has been studied through the medium of cold models by Saunders (61) for the blast furnace and by Hughes (42) for the gas producer. Segregation of the burden in a blast furnace causes uneven gas flow through the mass, the velocity being highest where the material is coarsest. Saunders studied the effect on segregation of the shape of the furnace walls, the size of the distribution bell at the top, and the shape of the throat round the bell. In was shown in Chap. 5 that the condition for kinematic similarity in the movement of granular solids is equality of Froude numbers with respect to solid flow. To ensure that friction and rebound effects are similar, the same materials should be charged to the model furnace as to the prototype, but with a particle size proportionate to the scale ratio. Irregularities of the furnace walls should be geometrically reproduced in the model. Where the upward flow of gases is rapid enough to have an appreciable effect upon the movement of particles, the air velocity in the model must be reduced in proportion to the square root of the linear scale ratio, thus maintaining equality of Froude numbers with respect to air flow. The condition for geometrically similar segregation of particles in model and prototype is then that the following groups shall be constant:

$$\frac{\rho_s h}{gd} \qquad \frac{\rho_g v_g{}^2}{gd'}$$

where ρ_s, ρ_g = density of solids, gases

h = height from which solids are dropped

d = mean particle diameter

d' = hydraulic diameter of voids between particles

v_g = mean velocity of gases

g = acceleration of gravity

For homologous systems where ρ_s and ρ_g are constant, the scale equations are

$$h = d = L \tag{16-5a}$$
$$v_g = \sqrt{d'} \tag{16-5b}$$

For cold models,
$$v = \sqrt{\frac{d'}{\rho_g}} \tag{16-5c}$$

The effect of segregation on the gas-flow distribution can be examined by the starch-iodine method of Bennett and Brown (27). They used a transparent slice model of a gas producer with broken glass representing the fuel. The glass was coated with a 20 per cent starch solution, and the air contained iodine vapor. The path of the main air stream was shown by a vein of color. Saunders (61) used a similar technique with slice models of blast furnaces. Hughes employed marble fragments coated with lead acetate and air containing sulfureted hydrogen.

Where air is introduced into a fuel bed through a high-velocity nozzle such as the tuyère of a blast furnace, the air stream completely changes the packing of the burning fuel by creating a hollow zone within which the particles exhibit a violent boiling movement. Models will not give a true picture of conditions in this zone unless the consumption of the solid particles by the fluid is simulated in the model. Newby (55) reproduced this boiling effect accurately by means of a cold-water model in which the fuel was replaced by crystals of sodium carbonate. With this model, he investigated problems of tuyère design.

Coming now to hot models, the simplest is the combustion pot, which can be treated as a full-scale element of a continuous furnace, for example, a chain-grate stoker (33). Given equal initial thicknesses of fuel bed and rates of air flow per square foot, the state of the fuel bed at different points along the chain grate is simulated by the condition in the pot at different times. The combustion pot has also been applied to the study of sprinkler and underfeed stoker fuel beds (52, 56) and to the behavior of a sinter strand for sintering iron ore mixed with coke (70). The method gives accurate results provided there are no major wall effects on packing or heat loss from the pot to vitiate similarity.

Heat Transfer

Heat transfer in furnaces may take place by conduction, natural convection, forced convection, and radiation. The generalized dimensionless equation for heat transfer is given in Chap. 7, Eq. 7-18. The criteria for thermal similarity in geometrically similar systems are:

Conduction and forced convection......... Peclet number
Natural convection...................... Grashof number
Radiation............................... Thring's radiation group

It was shown in Chap. 7 that in geometrically similar systems these three groups are incompatible, so that in general strict thermal similarity is not possible unless the effect of one of the above criteria greatly predominates over the effect of the other two.

The principal furnace heat-transfer phenomena in which conduction predominates are heat losses through furnace walls. The cross section of the conduction path is rarely constant, and the shape of the refractory walls is sometimes such that the differential form of Fourier's equation cannot be integrated mathematically. In this case, an electrical analogue model can be used to effect the integration. The integral form of Fourier's equation may be written

$$H = kF \, \Delta T$$

where H = total heat flow, Btu/hr
k = thermal conductivity
ΔT = temperature difference
F = a shape factor which is equal for conduction of heat and electricity.

The corresponding form of Ohm's law is

$$I = k'FE$$

where I = current flow
k' = electrical conductivity
E = potential difference

If, therefore, a scale model of the required refractory shape is made in a rather poor electrical conductor of known conductivity, the shape factor F may be determined by means of a battery, ammeter, and voltmeter connecting a pair of electrodes covering the internal and external isothermal surfaces.

A variation of the above method used by Langmuir, Adams, and Meickle (159) is to simulate the refractory shape by a hollow cell, the inner and outer heat-transfer surfaces being represented by copper plates and the other walls composed of nonconducting material such as glass. The cell is filled with a copper sulfate solution of known concentration and conductivity. A comparison of the electrical conductance of the cell with one of constant cross section containing the same electrolyte gives the shape factor directly. The current density should be low, to avoid polarization.

Often the cross section of the conductor is uniform at right angles to the direction of heat flow, in which case a two-dimensional model will give the required shape factor. Awberry and Schofield (146) used two-dimensional sections to scale on special conducting paper (resistance 2,500 ohms/in.^{2}) and outlined the isothermal surfaces in highly conducting silver paint. The resistance of the unknown shape was balanced on a Wheatstone bridge against a standard of known shape factor. The preferred standard shape was an annulus, since this shape is free from end effects. The shape factor for an annulus is

$$F = \frac{2\pi}{\ln (r_o/r_i)}$$

where r_o/r_i = outside/inside radius.

In the interior of a high-temperature furnace such as a blast furnace, heat transfer takes place by conduction, convection, and radiation in parallel. That is, heat reaches each particle of charge simultaneously by conduction from neighboring particles, convection from the hot gases, and radiation from the burning fuel. Conduction is rarely a controlling factor since most solid fuels and many process materials are poor conductors of heat. Convection also is relatively unimportant because it is directly proportional to the temperature difference, whereas radiation obeys the fourth power law. In most cases, therefore the predominant heat-transfer process is radiation. It was shown in Chap. 7 that, if the walls of the model are of such material and thickness that the heat loss per square foot by conduction is the same as in the prototype, then the radiation and Peclet groups are no longer incompatible and the scale equations for similarity with respect to radiation and conduction from the walls are (7-21a) to (7-21d). Such a model furnace should attain the same temperature profile as the prototype.

Equations (7-21a) to (7-21d) mean that for similarity with respect to radiation and conduction through the walls the heat input to the furnace should be proportional to the wall area, giving a constant heat release per square foot of wall. The heat release per cubic foot will then be higher in the model than in the prototype. This relation was tested experimentally by Thring (65), using a turbulent gas flame inside a vertical water-cooled cylinder. Two similar models were used, having a scale ratio of 3. Flame-radiation intensity and CO_2 concentration were measured at various heights up the cylinder. Figure 16-2a shows the radiation and CO_2 curves obtained with the prototype, Fig. 16-2*b* with the model at equal Reynolds number, and Fig 16-2*c* with the model at equal heat input per unit wall area, corresponding to constant radiation and modified Peclet number in Eq. (7-21). The curves of Fig. 16-2*c* are much more nearly similar to those of Fig. 16-2*a* than those of Fig. 16-2*b* at constant Reynolds number, showing that under these conditions radiation is the predominant process and forced convection can be neglected.

The heat-release pattern in a furnace may be determined from a cold model by the technique already described for finding the shape of the flame. A tracer gas is added to the nozzle fluid, and samples are withdrawn from the model furnace through a probe. To calculate the fraction of fuel energy which will be released by combustion at a given cross section of the furnace (a quantity which enters into the calculation of a heat balance in stages), it is necessary to multiply the concentration at each point by the corresponding velocity at that point. Leys (47) describes an intrument by which the gas can be simultaneously sampled and its velocity measured. The instrument gives the total head at the sampling point, and the static pressure is measured at the wall of the cross section.

A model flame-radiation furnace has been built at Ijmuiden in Holland for the study of problems connected with the open-hearth furnace (35 to 37). The linear scale ratio with respect to a large industrial furnace is about $L = 3$. It was felt that this was the greatest scale reduction that could give useful results in view of the incompatible similarity criteria involved. For example, Reynolds similarity would require a fuel input reduced in proportion to L, radiation-conduction similarity in proportion to L^2, equality of dwell time of fuel particles in the flame in proportion to L^3. The latter is probably the condition for equal formation of soot.

Leckie and his colleagues (44, 45) have made extensive use of a scale-model gas-fired open-hearth furnace to investigate the effect on thermal efficiency and heat distribution of port design, draft, air/gas ratio, and flame luminosity. In the model, the charge of steel was replaced by a sectional calorimeter covered with thin refractory tiles. The ratio of heat input radiated to the charge is given by the group*

$$\frac{H}{\sigma e L^2 (T_1{}^4 - T_2{}^4)}$$

The model furnace was operated without air or gas preheat and with a flame temperature of about 1000°C, as against 1600°C in a proto-

---

* See Eq. (5-38).

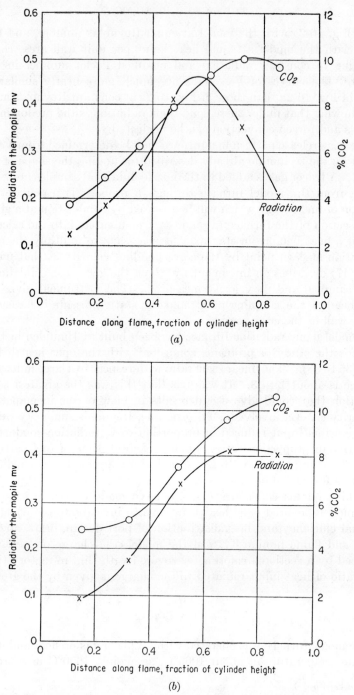

Fig. 16-2. Thermal similarity in water-cooled furnaces. [*M. W. Thring, Research,* flame. (*a*) Prototype. (*b*) Model (scale ratio **L** = 3) at same Reynolds number according to Eq. (7-21) and neglecting the Reynolds number.

type furnace. Radiation similarity was empirically achieved by varying the thickness of the calorimeter tiles until a fraction of heat given
up to the calorimeter in the model was the same as that given up to the
charge in the prototype. It was assumed that this tile thickness would
give approximately correct effects for change in design and operating
conditions. Owing to the absence of preheat in the model and the
much lower kinematic viscosity of the gas and air, it was possible to
operate the model at approximately the same Reynolds number in the
gas ports as the prototype without having excessive velocities on the
small scale. This method of empirical similarity is often useful where
theoretical similarity is not practicable.

Physical and Chemical Changes in the Charge

Physical Changes. The principal physical changes that occur in furnaces are the melting of solids and solidification of fused materials.

A warm model has been used to study the melting of steel scrap in
an open-hearth furnace (65). It was desired to observe the behavior
of the change during melting and compare rates of melting with different types of scrap and methods of charging. In the model, the steel
scrap was simulated by suitably shaped pieces of paraffin wax, and an

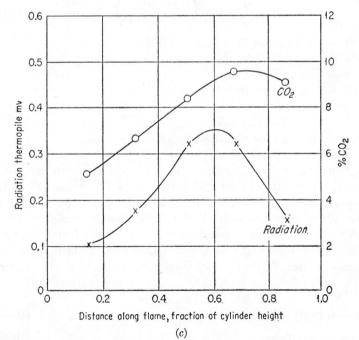

(c)

2: 36 (1949).] Radiation intensity and CO_2 concentration along the axis of the
and air/gas ratio as (a). (c) Model at same thermal similarity criterion as (a),

electric bar heater occupied a position corresponding to that of the flame. It was found empirically that, if the model heating element was supplied at about 500 watts and the bath heated for 10 min before charging, the behavior of paraffin-wax shapes simulated that of scrap steel in a full-sized furnace, 36 min in the model corresponding to about 8 hr in the prototype (t = 13.3).

With this model, it was found that various methods of charging had little effect in the melting time but that light scrap fused more quickly than heavy scrap and the time for complete melting could be reduced to one-seventh by stirring of the bath. Melting on a sloping shelf was also found to be an advantage. In carrying out these experiments, it was found necessary to correct for a steady change in the radiation from the model heating element which caused a corresponding change in the time scale. This work is another example of the application of empirical similarity.

Models have been employed for studying the continuous casting of metals (29, 258). Both warm models using paraffin wax and hot models using molten metal have been investigated. For geometrically similar systems, the dimensionless heat-transfer equation may be written

$$\frac{H}{kd\,\Delta T} = \phi\left(\frac{vc\rho d}{k}, \frac{vc\rho}{\sigma(\Delta T)^3}\right) \tag{16-9}$$

where H = total rate of heat flow = dQ/dt

k = thermal conductivity of mold wall

d = diameter of mold

ΔT = temperature difference

v = velocity at which the metal is being pulled out of the bottom of the mold

c, ρ = specific heat, density of metal

σ = Stefan-Boltzmann constant

The two groups on the right-hand side of Eq. (16-9) represent, respectively, the ratio of heat withdrawn by bodily movement of the cast metal from the mold to heat loss by conduction through the walls and the ratio of heat withdrawn bodily to heat loss by radiation across the gap between the frozen metal and the wall. The former group is likely to be controlling at low temperatures and the latter group at high temperatures. The two are incompatible as similarity criteria and give rise to different time scales. The scale equations are:

Conduction controlling: $v = \dfrac{1}{d}$ (16-9a)

Radiation controlling: $v = 1$ (16-9b)

The correct value of **v** is that at which the metal is frozen across the same fraction of its diameter at corresponding cross sections of the mold in both model and prototype.

Chemical Changes. It was shown in Chap. 7 that chemical reactions can be usefully scaled up only from homologous systems. This means that the small-scale investigation of chemical changes in furnaces must be conducted by means of hot models.

One of the most important industrial furnaces in which chemical changes occur is the blast furnace. Two different similarity criteria for blast furnaces have been proposed, depending upon the regime that is assumed to exist. Thring (29, 258) pointed out that, if diffusivity processes are rate-determining, the regime is dynamic and model and prototype should be compared at equal Reynolds numbers. Traustel (69) considered that what he termed the "Guldberg-Waage number" should be the same in model and prototype. The Guldberg-Waage number G_w is defined by the equation

$$G_w = V_1{}^{n_1} V_2{}^{n_2} \cdots - \frac{U_1{}^{m_1} U_2{}^{m_2} \cdots}{k_v}$$

where $U_1 U_2 \cdots$ = partial volumes of reacting gases
$m_1 m_2 \cdots$ = numbers of molecules of each reacting gas taking part in reaction
$V_1 V_2 \cdots$ = partial volumes of reaction products
$n_1 n_2 \cdots$ = numbers of molecules of each reaction product formed by reaction
$k_v \cdots$ = dimensionless partial equilibrium constant

Since there are several different reactions required to be similar, the necessary conditions for similarity under a chemical regime are:

1. Equal temperatures, gas compositions, and total gas pressures in model and prototype

2. An equal ratio of rate of gas flow to rate of formation of reaction products at the surface of the particles

3. An equal ratio of heat transport in the gas stream to convectional heat transfer from the particles

For a geometrically similar model in which the particle size of the furnace charge is scaled down in the same ratio as the furnace itself, the scale equation is given in Chap. 7,

$$\mathbf{v} = \mathbf{H} = 1 \tag{7-29a}$$

where v = mean linear velocity of gases.

If the particle size is not scaled down in the same ratio as the furnace, the scale equation for material flow (conditions 1 and 2) is

$$\frac{vd}{L} = 1 \qquad (16\text{-}10)$$

where d = average particle diameter
L = linear dimension of furnace
In this case, however, the third condition above is not satisfied.

If diffusion of oxygen from inside the lumps to the surface is of major importance in reducing iron ore, then there is a further similarity condition, which requires that in the model furnace the mean diameter of the ore particles shall be reduced in size by the square root of the scale factor for the fuel particles.

Bishop (28) describes an experimental blast furnace constructed to one-tenth scale ($L = 10$). The criterion of similarity was assumed to be the CO/CO_2 ratio in the exit gases, and the rate of blowing of the model was adjusted until the ratio was equal to that found on the full scale. This was found to occur when the linear gas velocity in the model was about one-third that in the prototype, suggesting the empirical scale equation

$$v = \sqrt{L} \qquad (16\text{-}11)$$

This is nearer to Eq. (7-29a) for a chemical regime than to Eq. (16-1) for a dynamic regime, but it is sufficiently different from either to indicate that more data are needed on the regime existing in blast furnaces.

Flow of Molten Materials and Slags

Pechès (57) has used models to examine the convection currents in molten glass in a glass tank. These currents are of great importance in glassmaking because, on the one hand, they erode the refractory walls and, on the other, they promote mixing of materials and the elimination of air bubbles. A glass tank usually consists of two chambers separated by a bridge wall under which the molten glass flows, the object being to have the glass fully molten and mixed before it passes under the wall to the second chamber, whence it is withdrawn. Convection currents along and across the chambers are superimposed on the over-all flow along the tank as glass is withdrawn.

The model tank was to one-tenth scale and constructed with transparent sides. An electric heater above the working end simulated the cross-fired flames of a real furnace. Glycerin working over the range of 70 to 20°C gave approximately the variation in viscosity corresponding to that of glass between the melting and working temperatures.

Polystyrols were also used. In the model, the vertical temperature gradient tended to be too great compared with the horizontal one, and it was necessary to insulate the base of the model well or even apply some heat. Flow patterns were observed by introducing a thin filament of colored glycerin. The conditions for dynamic similarity in such a model have been worked out by Schild (62).

Where molten material is flowing under the influence of gravity, the similarity criterion is the Froude number. The flow of molten steel running out of an open-hearth furnace and splashing on the bottom of a mold has been simulated by a cold-water model. An equal degree of splashing is obtained when the height from which the liquid is poured is scaled down in proportion to the dimensions of the model.

Flow of Cooling Water

A simple application of models is to study the flow pattern of cooling water in furnace parts of irregular shape, such as the tuyères or blast nozzle of a blast furnace or the doorframe of an open-hearth furnace. These metal parts are subject to very intense heating, sometimes by direct radiation from an effectively black-body surface at 1700°C. A heat flux at this rate can be removed only by very rapid forced convection to high-velocity cooling water inside all the exposed parts. The old method was to circulate a very great quantity of cooling water. In many cases, the over-all temperature rise was only about 1°C, whereas efficient cooling could be achieved with one-thirtieth of the quantity of water provided the velocities were high and uniform over the hot surfaces. Leigh (46) studied the cooling-water flow in transparent plastic models of such parts, preferably at a Reynolds number equal to that in the original system, but in any case sufficiently high to give full turbulence. The flow in the model was rendered visible by a narrow beam of strong light from a slit illuminating small air bubbles injected into the water. By rearranging the positions of inlet and exit pipes and introducing carrying pipes inside the water space, it was often possible to produce a much more uniform flow.

Behavior of Refractories

Furnace refractories are liable to fail where subjected to direct flame impingement. The sticky-dust technique for detecting flame impingement by a means of a cold model is described under Furnace Aerodynamics.

A hot model furnace operating over the same temperature range as the prototype and under conditions of radiation similarity ($\mathbf{v} = 1$)

enables refractories to be tested under conditions closely resembling those of full-scale service.

SYMBOLS IN CHAPTER 16

a = cross-sectional area of burner nozzle
c = specific heat
D = diameter of flame
d = diameter of burner nozzle
d = mean particle diameter
E = electrical potential difference
F = shape factor
G = jet momentum
g = acceleration of gravity
H = heat flow in unit time
h = height
I = electrical current
K = const
k = thermal conductivity
k' = electrical conductivity
k_v = dimensionless partial equilibrium constant
L = linear dimension
l = length of flame
M = molecular weight
M = mass-flow rate
m = number of reacting gas molecules
N = mass ratio of (air + fuel)/fuel
N_0 = stoichiometric mass ratio of (air + fuel)/fuel
n = number of molecules of reaction product
p = pressure
r = radius, radial distance from axis
r_e = external radius of flame = $D/2$
T = temperature
t = time
U = partial volume of reacting gas
u = local velocity
V = partial volume of reaction product
v = mean velocity
x = distance from burner nozzle along axis of flame
y = mole fraction of fuel gas in stoichiometric combustion mixture
α = ratio of moles in stoichiometric mixture after combustion to ratio before combustion
Δ = difference (as in Δp, ΔT)
η = r/x
μ = viscosity
ν = kinematic viscosity
ρ = density
σ = Stefan-Boltzmann constant

Subscripts

a = air
f = furnace
g = gas, gases
i = inside
l = liquid
m = minimum
o = outside
rx = at a point P at axial distance x and radial distance r from burner nozzle
s = surrounding fluid
s = solid
x = at axial distance x from burner nozzle
0 = at burner nozzle

CHAPTER 17

MISCELLANEOUS EQUIPMENT

The object of this chapter is to describe some model studies of miscellaneous devices and types of process plant which do not fall under any of the broad headings already dealt with. Four short studies are presented, relating, respectively, to:

Ball mills
Pressure-jet spray nozzles
Centrifugal-disk atomizers
Screw extruders

Ball Mills

A *ball mill* consists essentially of a revolving steel drum about half filled with steel balls, the length of the drum being approximately equal to its diameter. Where the length of the drum is several times the diameter, the mill is termed a *tube mill*. The grinding mechanism and scale-up principles are the same for both types.

Ball and tube mills are old established grinding devices which are still extensively used in ore reduction, cement manufacture, and other industries. They can grind material either wet or dry. There is a considerable literature on both the theoretical and the practical aspect of ball milling; yet it remains something of an art, and the design or selection of full-sized equipment is commonly based upon experiments with small-scale mills.

As early as 1904, Fischer (79) studied the motion of balls in specially constructed mills and observed the phenomenon of the critical speed. In 1919, Davis (78) gave an exhaustive mathematical treatment of the ball mill. Later, there were important theoretical and experimental studies by Coghill, Gow, and their associates (77, 80, 81). Michaelson (82) has described an empirical method of designing large mills from laboratory grinding tests, and Matz (252) has discussed the application of model theory to the ball mill.

The movement of balls in a ball mill is illustrated in Fig. 17-1. All the balls outside the dotted spiral surface at any moment are moving as if they were fixed to the drum. Those inside this surface are falling down to replenish the flow at the bottom, and it is this falling impact which causes the grinding. The balls thus move in closed curves having a central point which is at a distance R_s from the axis of the mill and situated on a radius making an angle θ with the vertical. When the mill is rotated very slowly, the charge of balls rises until its surface makes an angle θ_0 with the horizontal, this being the angle of repose. As the mill rotates faster, the angle θ increases steadily. Balls adjacent to the shell begin to be carried up and cascade back to the bottom of the mill. When the speed of rotation is such that centrifugal force on the outermost layer of balls just balances the force of gravity, these balls are carried

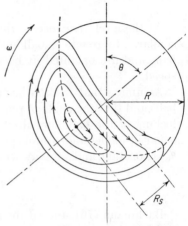

FIG. 17-1. Motion of balls in a ball mill.

right round without falling and the mill has reached its critical speed. If ω_c is the critical angular velocity in radians per second, m is the mass of a ball, r is the radius of a ball, and R is the internal radius of the mill, then

$$\omega_c^2(R - r)m = gm$$

or

$$\omega_c = \sqrt{\frac{g}{R - r}} \tag{17-1}$$

Assuming that r is small compared with R, then in engineering units the critical speed is given by

$$n_c = \frac{76.6}{\sqrt{D}} \tag{17-2a}$$

where n_c = critical speed, rpm, and D = inside diameter of mill, ft.

Neglecting r, Eq. (17-1) can be written

$$\frac{\omega^2 R}{g} = 1 \tag{17-2b}$$

The dimensionless group on the left is a Froude number and constitutes the similarity criterion for the flow pattern of the balls. This is to be expected, since the Froude number represents the ratio of inertial to gravitational forces.   The performances of different mills should be compared at equal Froude numbers, i.e., at equal fractions of the critical speeds.   Corresponding speeds are given by the scale equation

$$n = \frac{1}{\sqrt{D}} \qquad (17\text{-}2c)$$

At these relative speeds, the motions of the balls are kinematically similar.   In practice, ball mills are generally about half filled with balls and are operated at between 65 and 85 per cent of the critical speed.

Assuming that friction has a negligible effect on the motion of the balls and that the angle of repose is a consequence of geometrical interlocking of the balls, or, alternatively, that coefficients of friction are equal in model and prototype, then for geometrically similar systems θ_0 is constant and we may write

$$\theta = \phi\left(\frac{n}{n_c}\right) \qquad (17\text{-}3)$$

Bachmann (76) studied the motion of balls in a mill with glass ends and found empirically that

$$\theta - \theta_0 = \frac{\omega}{\omega_c} \qquad \text{radians}$$

or $\qquad\qquad \theta - \theta_0 = 57.3\,\frac{n}{n_c} \qquad \text{deg} \qquad (17\text{-}3a)$

Matz (252) has tried to derive this relation theoretically but his derivation goes beyond the bounds of similarity theory.   The results of Bachmann do, however, confirm the general validity of the Froude number as a similarity criterion.

It follows from kinematic similarity that at corresponding speeds the distance R_s of the center of motion from the axis of the mill will obey the relation

$$\frac{R_s}{R} = \phi\left(\frac{n}{n_c}\right) \qquad (17\text{-}4)$$

So far, only the motions of the balls and drum of the mill have been considered, without reference to the material being ground.   As long as the latter is a dry, nonsticky material and its mass is small in relation to that of the balls, it will not alter the similarity of flow conditions, since it merely causes a part of the inertial energy of the balls to be absorbed by impact with the material instead of with other balls. The power consumption is also unaffected, since the main energy is consumed in raising the balls against gravity.   Under these conditions, the power required to run a mill at a given speed is relatively constant, but the fineness of product varies with the duration of grinding or, in the case of continuous mills, the rate at which the material is fed. For dry grinding in mills of similar types run at corresponding speeds, the energy consumption per ton for a given degree of reduction of a given material tends to be constant.

Assuming that the power required to run a ball mill is the power needed to lift and cascade the balls, it will be proportional to the total mass of balls, the average height to which they are lifted, and the number of times this is repeated per minute. The latter is equal to the rpm, and for equal corresponding speeds the average height of lift is a constant fraction of the mill diameter. Hence, at equal corresponding speeds

$$P = C\rho'fLD^3n \qquad (17\text{-}5)$$

where P = power consumption,
ρ' = bulk density of balls
f = fraction of mill volume filled by balls
L = length of shell
D = inside diameter of shell
n = rpm
C = a const

Note that the power consumption depends not upon the size of the balls but only on their total volume and bulk density. Since at corresponding speeds n varies as $D^{-0.5}$, for mills of generally similar type in which ρ' and f are equal, we may write the scale equation

$$\mathbf{P = LD^{2.5}} \qquad (17\text{-}5a)$$

For geometrically similar mills, $\mathbf{D = L}$, and Eq. (17-5a) becomes

$$\mathbf{P = L^{3.5}} \qquad (17\text{-}5b)$$

Gow, Campbell, and Coghill (80) carried out an experimental study of the fundamentals of ball milling, using a series of mills of different diameters. As a measure of the amount of grinding, they took the tonnage throughput multiplied by the increase in surface per ton, mill performance being expressed as surface tons per horsepower-hour. Their conclusions were that for the same input material at equal percentages of the critical speed:

1. Horsepower varies as $LD^{2.6}$.
2. Mill throughput varies as LD^2.
3. Fresh surface created per ton of throughput varies as $D^{0.6}$.

4. Hence, performance expressed as surface tons per horsepower-hour remains constant.

Conclusion 1 is a reasonably good experimental confirmation of Eq. (17-5a). Conclusions 2 and 3, respectively, yield the scale equations

$$W = LD^2 \qquad (17\text{-}6a)$$

and

$$\delta = \frac{1}{D^{0.6}} \qquad (17\text{-}6b)$$

where W = mass rate of throughput, tons/hr, δ = mean particle diameter of product (inversely proportional to the surface per ton).

Equation (17-6a) states in effect that the residence time of material passing through a continuous mill should be constant.

In the modified horsepower equation quoted below [(17-8)], the exponent on D is changed to the theoretical value of 2.5. To conform with this and Eq. (17-6a), Eq. (17-6b) should become

$$\delta = \frac{1}{\sqrt{D}} \qquad (17\text{-}7)$$

A later paper by Gow, Guggenheim, Campbell, and Coghill (81) qualified conclusion 1 above, pointing out that in ball mills, where the length is short, end effects have an appreciable influence on power consumption. They proposed the following empirical formula for power consumption:

$$P' = \left[\left(\frac{L'}{2} - 1 \right) C' + 1 \right] \left(\frac{D'}{2} \right)^{2.5} P \qquad (17\text{-}8)$$

where P' = power consumed by a large ball mill of diameter D' ft and length L' ft

P = power consumed by a 2- by 2-ft laboratory mill operated on the same material with the same percentage ball milling and corresponding speed

C' = a const 0.9 where L' is less than 5 ft, 0.85 where it is greater

From this can be derived the scale equation

$$P = (0.9L + 0.1)D^{2.5} \qquad (17\text{-}8a)$$

where $L < 2.5$, and

$$P = (0.85L + 0.15)D^{2.5} \qquad (17\text{-}8b)$$

where $L > 2.5$.

These may be regarded as empirical modifications of Eq. (17-5a) for systems that are not geometrically similar.

Ball size does not enter into any of the above relations and makes relatively little difference to the power consumption and capacity-speed relation. Larger balls are not necessary for harder ores. Coghill and DeVaney (77) concluded that for producing closely sized particles the optimum ball size varies as the square root of the mean particle diameter, or

$$d^2 = C''\delta \qquad (17\text{-}9)$$

where d = optimum ball diameter
  δ = mean particle diameter
  C'' = a const varying with the material ground

Since larger mills tend to produce a finer product at the same corresponding speed, this relation suggests that they should contain smaller balls. Combined with (17-7), it gives the scale equation

$$d = \frac{1}{D^{0.25}} \qquad (17\text{-}10)$$

The known scale-up relations for all tube mills are thus partly theoretical and partly empirical. The case of complete geometrical similarity, with ball diameter and mean particle diameter of product scaled down in proportion to mill dimensions, is never met with in practice. On the contrary, the optimum ball size is larger and the product tends to be coarser on the small scale. On the basis of the above short review, the rules for scaling up may be summarized as follows:

1. Percentage of the mill volume filled with balls equal on the small and large scale

2. Bulk densities of random-packed balls equal in both cases (if the balls in each mill are of varying diameters this calls for a similar size distribution)

3. Relative diameters of balls in accordance with Eq. (17-10) (it will make little difference if the sizes are equal)

4. Speeds of rotation in accordance with Eq. (17-2c)

5. Mean particle size of feed to the mills equal in both cases

6. Mill throughputs in accordance with Eq. (17-6a)

These are the independent variables. The rules for the prediction of dependent variables under the above conditions are:

7. Power consumption given by Eq. (17-5b) where shells are geometrically similar or by Eq. (17-8a) or (17-8b) where they are not.

8. Mean particle size of product given by Eq. (17-7).

The above rules apply primarily to the dry milling of nonsticky materials. They may also apply to wet milling provided that the viscosity of the slurry is low and the solids do not form sticky aggregates. Otherwise, it would be safer to apply the indirect scale-up procedure proposed by Michaelson (82). This procedure is essentially to prepare a graded series of standard reference samples, the power requirements for which, in surface tons per horsepower-hour, are known for large-scale operations. The quantity of material through a given sieve mesh produced by a fixed number of revolutions in a laboratory mill is compared for the unknown sample and the reference samples, and the large-scale power requirements are interpolated from the known requirements for the two reference samples nearest in laboratory-scale grindability to the unknown sample. Empirical factors are used to correct for operating conditions different from those for which the reference-sample power requirements were determined.

Pressure-jet Spray Nozzles

The advent of the gas turbine with fuel-spray combustion chambers has led to a considerable amount of investigation in the operation of centrifugal spray nozzles which utilize the pressure energy of the fluid

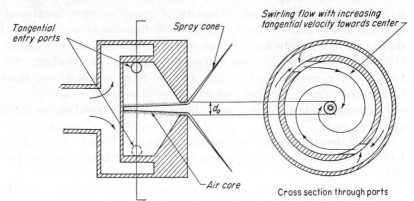

FIG. 17-2. Diagram of a pressure-jet spray nozzle.

to break it up into fine droplets. It is usual to make nozzles for this purpose geometrically similar to one another, and it is, therefore, useful to apply similarity theory to determine how far the flow properties and droplet size can be used to determine the corresponding properties for a nozzle of different size. The design of a simple nozzle of this type is shown in Fig. 17-2. The fluid is pumped at a high pressure through two tangential ports into the swirl chamber. If the viscosity of the fluid is low, then as it spirals into the center the angular momentum is conserved and thus the tangential velocity goes up inversely as the radius falls until eventually a point is reached where there is an air core because all the pressure has been converted to flow velocity.

The diameter of this air core is less than the diameter of the exit nozzle, and so the fluid immediately around the core spills out over the rim of the nozzle orifice with a combination of high tangential velocity and a certain axial velocity due to the residual static pressure at points outside the core. The part of the fluid which flows out at the rim of the orifice has the greatest axial velocity and the least tangential velocity; the part at the core has practically no axial velocity because the static pressure has fallen to atmospheric, but it has the greatest tangential velocity. Hence, the fluid issues as a conical sheet swirling round and with a tendency for the inner part to fly through the outer part. This sheet rapidly breaks up, mainly under the influence of its own turbulent motion rather than as a consequence of the interaction of the sheet with the surrounding atmosphere. The actual size of droplet formed must depend to some extent on surface tension, since the amount of energy used in breaking the liquid up into fine droplets depends on the area of new surface created. Actually, however, the surface tension of different liquid fuels only varies from the 21 dynes/cm for aviation gasoline to 30 dynes/cm for heavy fuel oil, whereas the kinematic viscosity at constant temperature varies from 0.62 to 1,200 centistokes for the same fuels. For this reason, experimental data are not available to determine how far the particle size depends on surface-tension forces. The size does depend strongly on viscous forces, because the viscous drag within the swirl chamber tends to destroy the conservation of angular momentum and thus greatly to reduce the tangential velocity of the stream coming out of the orifice. In fact, at a fixed supply pressure, an increase in the viscosity of the liquid actually increases the flow rate because it reduces the size of the hollow core and thus causes the orifice to run fuller.

Flow through a pressure-jet spray nozzle obeys the generalized dimensionless equation for a viscosity-controlled dynamic regime,

$$\frac{\Delta p}{\rho v^2} = \phi \left(\frac{\rho v d_o}{\mu} \right) \qquad (17\text{-}11)$$

or
$$\frac{\Delta p d_o^4}{\rho q^2} = \phi \left(\frac{\rho q}{\mu d_o} \right) \qquad (17\text{-}11a)$$

where Δp = pressure drop through the system

v = mean liquid velocity through the orifice calculated as if the orifice were running full

d_o = diameter of orifice

ρ, μ = density, viscosity of liquid

q = volumetric rate of flow

It has been found experimentally (112) that

$$\frac{\Delta p d_o{}^4}{\rho q^2} = C \left(\frac{\rho q}{\mu d_o}\right)^{0.22} \tag{17-12}$$

(C is a dimensionless constant).

Hence, for given values of d_o and q, $\Delta h \nu^{0.22} = \text{const}$, where Δh is the pressure expressed as head of liquid $= \Delta p/\rho$ and ν is the kinematic viscosity $= \mu/\rho$. It follows that an increase in kinematic viscosity reduces the head required to maintain the given rate of flow, a result contrary to that for normal flow in tubes. This is because with higher viscosity the swirl is reduced and the orifice runs fuller.

Equation (17-12) applies to all geometrically similar spray nozzles whatever the liquid, provided they are operated at a sufficient pressure to create a hollow core in the orifice stream. The simple power function of the Reynolds number is an approximation to a rather more complex variation, and the exponent changes slightly with the nozzle geometry.

Experimental measurements for droplet size may be correlated in terms of the Reynolds and Weber numbers, since both viscosity and surface tension enter.

Putting

d_p = Sauter mean diameter, i.e., the diameter of a particle having the same ratio of surface to volume as the actual distribution of droplets observed

σ = surface tension

and expressing the Reynolds and Weber numbers in terms of q rather than v, we may write the dimensionless equation

$$\frac{d_p}{d_o} = \phi \left(\frac{\rho q}{\mu d_o}, \frac{\rho q^2}{\sigma d_o{}^3}\right) \tag{17-13}$$

For a range of geometrically similar nozzles operated on the same fluid, the experimentally observed relationship (109) is

$$d_p = C' \frac{q^{0.32}}{\Delta p^{0.53}} \tag{17-14}$$

(The constant C' is not dimensionless but embodies density, viscosity, and surface-tension functions.)

Evaluating Δp from Eq. 17-12 in terms of q and d_o and substituting in Eq. (17-14),

$$\frac{d_p}{d_o} = C'' \frac{d_o{}^{1.24}}{q^{0.86}} \tag{17-15}$$

Using the empirical equation (17-15) to determine the function ϕ in Eq. (17-13), we obtain the dimensionless relation

$$\frac{d_p}{d_o} = C''' \left(\frac{\rho q}{\mu d_o} \right)^{-0.1} \left(\frac{\rho q^2}{\sigma d_o{}^3} \right)^{-0.38} \tag{17-16}$$

For homologous systems this reduces to the general scale equation

$$\mathbf{d}_p = \frac{\mathbf{d}_o{}^{2.24}}{\mathbf{q}^{0.86}} \tag{17-16a}$$

or, in terms of pressure drop,

$$\mathbf{d}_p = \frac{\mathbf{d}_o{}^{0.61}}{\mathbf{\Delta p}^{0.39}} \tag{17-16b}$$

(boldface type as usual representing ratios of quantities for prototype/model).

For the droplet diameter to be changed in the same ratio as the orifice diameter so that geometrical similarity is preserved and $\mathbf{d}_p/\mathbf{d}_o = 1$, the scale equations are

$$\mathbf{\Delta p} = \frac{1}{\mathbf{d}_o} = \frac{1}{\mathbf{L}} \tag{17-17a}$$

and

$$\mathbf{q} = \mathbf{d}_o{}^{1.44} = \mathbf{L}^{1.44} \tag{17-17b}$$

Equation (17-17a) represents the condition for dynamic similarity. Equations (17-16a) and (17-16b) permit extrapolation for identical liquids in geometrically similar nozzles at unequal Reynolds numbers.

It is possible that the ratio of the liquid viscosity to the viscosity of the air or gas into which it is sprayed (μ_a) should enter into Eq. (17-16) as an additional dimensionless term $(\mu/\mu_a)^\delta$. The exponent δ has not been determined.

Centrifugal-disk Atomizers

Centrifugal-disk, or cup, atomizers are used in spray driers, oil burners, and humidifying equipment. Compared with pressure-jet and steam- or air-jet atomizers, they have the advantage that they cannot readily plug, are not appreciably affected by erosion, and require only a low-pressure fluid feed since the work is done by the rotating disk.

The scale-up law depends upon the mechanism by which rupture of the sheet of fluid into droplets occurs. There are three possible mechanisms:

1. Individual drops break away from the disk as centrifugal force overcomes surface tension, as when liquid issues from a single small nozzle.

2. The liquid leaves the disk as a continuous sheet, which breaks up owing to friction or impact against the air.

3. The liquid leaves the disk as a continuous sheet, which breaks up as a result of its own internal turbulence.

1. The first mechanism certainly occurs when the feed rate to the disk is kept so low that the liquid flows in separate spirals to the rim. In this case, the droplets consist chiefly of uniform large ones, with much smaller satellites between them.

Walton and Prewett (117) showed that under these circumstances liquid accumulates at the rim of the disk until the ratio of the product of drop mass and acceleration to the product of surface tension and linear dimension reaches a certain fixed value.   Thus,

$$\frac{(\pi d_p{}^3 \rho / 6)(n^2 D/2)}{\sigma d_p} = \text{const}$$

where D = diameter of disk

d_p = Sauter mean diameter of drops (see Pressure-jet Spray Nozzles)

ρ, σ = density, surface tension of liquid

n = rotational speed of disk, revolutions in unit time

whence the general dimensionless relation

$$\frac{d_p}{D} = \phi\left(\frac{\rho D^3 n^2}{\sigma}\right) \tag{17-18}$$

The group in parentheses is a Weber number based on the disk radius and peripheral speed.

2. In the second type of mechanism, where the sheet breaks up as a result of impact with the air, there will be a maximum size of droplet which is just stable under the opposition of the inertial resistance of the air and the surface tension holding the droplet together, so that

$$\frac{\rho_a v^2 d_p{}^2}{\sigma d_p} = \text{const}$$

where v = flying velocity of droplets through air and ρ_a = density of air.

This gives a critical Weber number for maximum droplet size,

$$\frac{\rho_a v^2 d_p}{\sigma} = \text{const}$$

or, in general,

$$\frac{d_p}{D} = \phi\left(\frac{\rho_a v^2 d_p}{\sigma}\right)$$

Assuming that v is proportional to the peripheral velocity of the disk, $\pi n D$,

$$\frac{d_p}{D} = \phi\left(\frac{\rho_a n^2 D^3}{\sigma}\right)$$

$$= \phi\left(\frac{\rho n^2 D^3}{\sigma} \frac{\rho_a}{\rho}\right) \tag{17-19}$$

3. Third, when the sheet breaks up as a result of its own internal turbulence, the droplet size is determined by the ratio of internal shearing forces in the sheet to the surface tension forces in the droplets. Thus,

$$\frac{\mu(v/\delta^2)\delta^3}{\sigma d_p} = \text{const}$$

or $$\frac{\mu v \delta}{\sigma d_p} = \text{const} \tag{17-20}$$

where δ = thickness of sheet at rim of disk and μ = viscosity of liquid.

v is proportional to the peripheral speed $\pi n D$, and δ is proportional to $q/vD = q/\pi n D^2$, where q = volumetric rate of feed of liquid. Hence, (17-20) becomes

$$\frac{\mu q}{\sigma D d_p} = \text{const}$$

or generally $$\frac{d_p}{D} = \phi\left(\frac{\mu q}{\sigma D^2}\right)$$

$$= \phi\left(\frac{\mu D}{\rho q} \frac{\rho q^2}{\sigma D^3}\right) \tag{17-21}$$

The first group on the right hand side of Eq. (17-21) is an inverted Reynolds number, and the second is the Weber number expressed in terms of volumetric rate of feed.

Assuming that all three of the mechanisms discussed above are operative, the complete generalized dimensionless equation for the centrifugal-disk atomizer should have the form

$$\frac{d_p}{D} = \phi\left(\frac{\rho D^3 n^2}{\sigma}, \frac{\rho q^2}{\sigma D^3}, \frac{\mu D}{\rho q}, \frac{\rho}{\rho_a}\right)$$

Instead of showing two Weber numbers, one can replace one of them by their ratio = $q^2/n^2 D^6$ or, more simply, by the square root of this group. Then,

$$\frac{d_p}{D} = \phi\left(\frac{\rho q^2}{\sigma D^3}, \frac{q}{n D^3}, \frac{\rho q}{\mu D}, \frac{\rho}{\rho_a}\right) \tag{17-22}$$

Friedman, Gluckert, and Marshall (104) carried out a theoretical and experimental study of the centrifugal-disk atomizer. They applied dimensional analysis to the factors discussed above, but omitting the density of the air, and derived a relation which for geometrically similar systems may be written

$$\frac{d_p}{D} = C \left(\frac{\mu D}{\rho q}\right)^{\alpha} \left(\frac{q}{nD^3}\right)^{\beta} \left(\frac{\rho q^2}{\sigma D^3}\right)^{\gamma} \tag{17-23}$$

The values of the exponents deduced from the experimental results were

$$\alpha = -0.2$$
$$\beta = 0.6$$
$$\gamma = -0.1$$

Equations (17-22) and (17-23) show that the centrifugal-disk atomizer is subject to a mixed regime. For homologous systems, the Reynolds and Weber numbers call for incompatible relations between q and D. Hence, dynamic similarity is not possible between disks of different sizes atomizing the same liquid, and experiments are preferably made with a full-scale system. Over a limited range of disk diameters, rotational speeds, and feed rates, the Sauter mean drop diameter may reasonably be scaled up or extrapolated by means of an empirical scale equation derived from (17-23),

$$d_p = \left(\frac{q}{n}\right)^{0.6} \frac{1}{D^{0.7}} \tag{17-23a}$$

It is probably within the experimental error to put

$$d_p = \left(\frac{q}{nD}\right)^{0.6} \tag{17-23b}$$

Screw Extruders

The operation of extrusion consists in forcing a plastic solid through a die so that it emerges as a continuous rod or tube of the desired cross section. It is an important operation in the plastics industry and is also employed in the manufacture of explosives (e.g., cordite).

Extrusion machines are of two main types: batch extrusion presses and continuous screw extruders. The press was the earlier type of extrusion machine, but except for some special applications it is tending to be displaced by the screw extruder with its continuous operation and greater output per die. A screw extruder consists essentially of a shaft and helix rotating in a stationary barrel. At one end of the barrel is the die through which the material is forced; near the other end is the feed hopper. The diameter of the shaft and the angle of the helix may vary along the length of the extruder.

The flow pattern in a screw extruder is complex, and the calculation of performance is complicated by the generation of heat, which is greatest where the shear is greatest and causes a nonuniform softening of the plastic. Design equations have been worked out for the ideal case of isothermal extrusion of a Newtonian fluid, but for the adiabatic compression, melting, and extrusion of a granular charge experimental data from a model extruder constitute the only reliable basis for design.

The flow in a screw extruder may be considered in two stages, flow along the barrel caused by the screw, leading to a build-up of pressure behind the extrusion die, and flow through the die under the influence of the pressure built up in the barrel. Flow through the die must equal the net flow along the barrel, and by equating the expressions for the two stages the model laws can be derived.

Flow through the die is relatively simple. It is a case of streamline flow through a nozzle and obeys Poiseuille's equations, which for geometrically similar systems may be written

$$\frac{q}{L^3} \propto \frac{\Delta p}{\mu} \tag{17-24}$$

where q = volumetric rate of flow

L = any linear dimension

Δp = pressure drop across die

μ = viscosity of extruded material or, in the case of non-Newtonian fluids, apparent viscosity

(In deriving scale relations for geometrically similar systems, it is not necessary to evaluate numerical constants or shape factors, since they disappear in the final scale equations. To avoid a multiplicity of constants, the "varies with" sign $\propto$ is often used instead of the equality sign =.)

Conditions in the barrel of a screw extruder are more complex. Carley and Strub (131) have described the various types of flow that take place. These are more easily visualized if the screw is imagined to be stationary while the barrel rotates. The velocity of any point on the barrel may then be resolved into two components, one parallel to the flights of the helix and one at right angles to them. The parallel component drags material along the helical channel toward the extrusion die. This is called *drag flow*. The component at right angles sets up a circulation of material across the channel, called *transverse flow*. The back pressure created by the die causes a certain amount of back flow along the bottom of the helical channel adjacent to the shaft, called the *pressure flow*, and also some leakage back through the clearance between the helix and the barrel, called *leakage flow*. The overall result is that material adjacent to the barrel is moved forward toward the die at a rate that is slightly diminished by back leakage across the flights, while there is a slow movement away from the die near the shaft of the helix.

If the volumetric drag flow, pressure flow, and leakage flow are, respectively, denoted by q_d, q_p, and q_l, the net forward flow in unit time q is given by

$$q = q_d - (q_p + q_l) \qquad (17\text{-}25)$$

The transverse flow absorbs power and effects some mixing of the material but does not contribute to the forward flow. If the die is removed, the back pressure is reduced to practically zero, pressure and leakage flow disappear, and $q = q_d$.

Carley and Strub (131) give both differential and integral equations for the three types of flow under isothermal conditions. The integral equation for drag flow is

$$q_d = \pi D n b^2 \cos^3 \varphi F_d$$

where D = mean diameter of screw
$\quad n$ = rotational speed of screw,
$\quad b$ = length between flights parallel to screw axis
$\quad \varphi$ = angle between helix and circumference of barrel
$\quad F_d$ = shape factor for drag flow
For pressure flow, the corresponding equation is

$$q_p = \frac{b h^3 \cos \varphi}{\mu F_p} \frac{dp'}{dz}$$

where h = thread depth
$\quad p'$ = pressure in barrel
$\quad z$ = length of helical channel
$\quad F_p$ = shape factor for pressure flow
For leakage flow,

$$q_l = \frac{\pi D E \delta^3 \, \Delta p'}{12 e \mu}$$

where E = a correction factor for eccentricity of screw
$\quad \delta$ = clearance between top of flights and barrel
$\quad e$ = width of lands measured in direction of screw axis
$\quad \Delta p'$ = total rise of pressure in barrel
Of the quantities entering into the above equations, the angle φ and the ratios between the linear dimensions b, D, e, h, z, and δ depend

only upon the geometry of the system.   Hence, for geometrically similar extruders one may write

$$q_d \propto nL^3$$

$$q_p \propto \frac{\Delta p'}{\mu}$$

$$q_l \propto \frac{\Delta p'}{\mu}L^3$$

where L = any linear dimension.

Hence, for geometrically similar extruders, Eq. (17-25) becomes

$$\frac{q}{L^3} = C_1 n - C_2 \frac{\Delta p'}{\mu} \qquad (17\text{-}26)$$

where C_1, C_2 are constants.

Assuming that the material is at the same absolute pressure (usually atmospheric) at both the feed to the extruder and the outlet from the die, then $\Delta p'$ for the barrel equals Δp for the die.   From Eqs. (17-24) and (17-26),

$$\Delta p \propto n\mu \qquad (17\text{-}27)$$

$$\frac{q}{L^3} \propto n \qquad (17\text{-}28)$$

Where the rotational speeds of model and prototype screw extruders are equal, that is, $\mathbf{n} = 1$, then in homologous systems where μ is constant the pressures generated by model and prototype are equal, i.e.,

$$\mathbf{\Delta p} = 1 \qquad (17\text{-}29a)$$

Volumetric extrusion rates are proportional to extruder volumes, i.e.,

$$\mathbf{q} = \mathbf{L^3} \qquad (17\text{-}29b)$$

Power consumption P varies as pressure rise multiplied by volumetric extrusion rate, and since Δp is constant,

$$\mathbf{P} = \mathbf{L^3} \qquad (17\text{-}29c)$$

The above scale relations were deduced by Carley and McKelvey (130).

For perfectly isothermal extrusion of the same material however, μ is unity irrespective of the speed, and geometrically similar extruders are kinematically similar at any speeds.   In adiabatic extrusion, it is necessary to have equal rates of shear in order that changes in temperature and viscosity at corresponding points in the model and prototype shall be the same.   Equal rates of shear are given by equal rotational speeds, and hence a necessary condition for similarity in homologous adiabatic extrusion systems is that

$$\mathbf{n} = 1$$

in which case Eqs. (17-29a) to (17-29c) apply. This condition is necessary also where conditions are intermediate between isothermal and adiabatic, i.e., in all practical cases.

Whether extrusion is approximately adiabatic or whether the barrel is externally heated, it is necessary that the surface-heat loss or gain in the model per unit of throughput be adjusted by means of appropriate jacketing to equal that in the prototype. This aspect of thermal similarity was discussed in Chap. 9.

Carley and McKelvey (130) compared the performance of a 20-in.-diameter plasticizing extruder operating almost adiabatically and a 2-in.-diameter model ($L = 10$). The barrel of the model was insulated and wired to balance heat losses, control being maintained by measuring at many points the temperature gradient in the barrel wall. These gradients were adjusted to zero before performance data were measured. Table 17-1, adapted from Carley and McKelvey's paper, shows how the scaled-up model performance compared with full-scale measurements. The higher exit temperature from the model may

TABLE 17-1. SCALE-UP RESULTS FOR ADIABATIC SCREW EXTRUDERS*

	Model (experimental results)	Prototype	
		Calculated from model	Observed
Diameter, in.	2	20	20
Speed, rpm	15	15	15
Capacity, lb/hr	2.5	2,500	2,400
Power, hp	0.29	290	245
Pressure, psi	1,850	1,850	1,600
Exit temperature, °C	169	169	156

* J. F. Carley and J. M. McKelvey, *Ind. Eng. Chem.*, **45**: 985 (1953).

have been due to the higher power consumption, or to the fact that the full scale extruder was not perfectly adiabatic. On the whole, however, agreement is amply good enough for design purposes, and such data could not have been obtained by calculation alone.

SYMBOLS IN CHAPTER 17

b = length between screw flights parallel to axis.

$C, C', C'', C''', C_1, C_2$ = const

D = diameter

d_0 = diameter of orifice

d_p = Sauter mean diameter of droplets
E = correction factor for eccentricity of screw
e = width of lands in direction of screw axis
F_d = shape factor for drag flow
F_p = shape factor for pressure flow
f = fraction of ball-mill volume filled by balls
g = acceleration of gravity
h = thread depth
Δh = increase in head
L = length
m = mass
n = rotational speed, revolutions per unit time
n_c = critical rotational speed
P = power consumption
P' = power consumed by a ball mill of internal diameter D'
     and length L'
p' = pressure in extruder barrel
$\Delta p'$ = pressure rise in extruder barrel
Δp = pressure difference
q = (net) volumetric rate of flow
q_d = rate of drag flow
q_l = rate of leakage flow
q_p = rate of pressure flow
R = internal radius of ball mill
R_s = distance of center of motion of balls from axis of mill
r = radius of ball
v = fluid velocity
z = length of helical channel
α, β, γ = exponents
δ = exponent
δ = mean particle diameter of milled product
δ = thickness of liquid sheet
δ = clearance between screw flights and barrel
θ = angle
θ_0 = angle of repose
μ = viscosity
μ_0 = viscosity of air
ν = kinematic viscosity = μ/ρ
ρ = density
ρ' = bulk density of balls
ρ_a = density of air
σ = surface tension
φ = angle between helix and circumference of barrel
ω = angular velocity, radians/sec
ω_c = critical angular velocity

CHAPTER 18

CORROSION

It is notoriously difficult to predict the degree of corrosion of chemical plant in service by means of laboratory tests. One reason is that the rate of attack depends not only on the chemical composition and temperature of the process fluids and materials of construction but also on the fluid velocity, flow pattern, and geometry of the system. It is an important function of the pilot plant to furnish more reliable information about corrosion under service conditions than can be obtained in the laboratory.

In some systems, the rate of corrosion is actually controlled by the fluid velocity, in which case discrepancies between laboratory corrosion tests and plant results are particularly to be expected. The existance of a dynamic or a mixed dynamic-chemical regime is shown, as described in Chap. 6, by the relatively low temperature coefficient of the corrosion rate and its dependence on the rate of stirring (Reynolds index).

The factors which favor a dynamic regime in corrosion systems are as follows:

1. A slow rate of diffusion of the corroding agent toward the metal surface or of the corrosion products away from it

2. Concentration differences in an electrolyte, giving rise to galvanic currents and electrochemical attack

3. Depletion of metal ions adjacent to the surface by a rapidly moving electrolyte, also causing galvanic currents

4. Erosion by suspended matter, solid, liquid, or gaseous

Diffusion Control

The materials of construction of chemical plant are normally chosen because their rate of corrosion is low, and this rate is not likely to be diffusion-controlled unless the corroding agent is in a very dilute state, e.g., sulfur dioxide in air or oxygen dissolved in water. In such cases, there may be a dynamic or mixed regime. Under a dynamic regime, the similarity criterion is the Reynolds number. For simple diffusion of a soluble constituent toward or away from a solid surface, the mass-transfer coefficient at equal Reynolds numbers varies inversely as the linear dimension, as was shown in Chap. 7 [Eq. (7-15b)].   In so far as corrosion is diffusion-controlled, therefore, the product $L\delta$ for homologous systems should be constant at equal Reynolds numbers, where L is a characteristic linear dimension (e.g., pipe diameter) and δ is the average rate of penetration measured at corresponding points.

Speller and Kendall (75) investigated the corrosion of iron pipes of different diameters by Pittsburgh tap water at various velocities and temperatures.   The corrosion was attributed to dissolved oxygen, and its low temperature coefficient and marked velocity dependence would

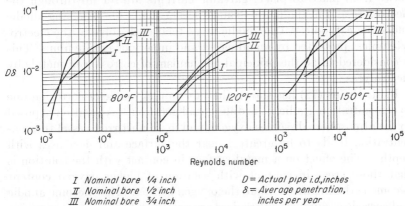

I   Nominal bore ¼ inch D = Actual pipe i.d.,inches
II Nominal bore ½ inch δ = Average penetration,
III Nominal bore ¾ inch   inches per year

FIG. 18-1. Initial corrosion rates of iron pipes by tap water at 80, 120, and 150°F. [*F. N. Speller and V. V. Kendall, Ind. Eng. Chem.*, **15**: 134 (1923).]

suggest a dynamic regime.   The diffusion mechanism is here complicated by the formation of a solid corrosion product, ferric hydroxide, which partly protects the surface from further attack by creating an "effective film thickness" greater than that obtaining at the same water velocity over bare metal.   Figure 18-1 shows corrosion curves for water at 80, 120, and 150°F, flowing in pipes of ¼ to ¾ in. nominal bore, the results being here replotted as $D\delta$ versus Reynolds number. D is the actual pipe diameter in inches, and δ is the average rate of penetration in inches per year.   There is a general tendency for $D\delta$ in pipes of different sizes to be of the same order of magnitude at equal Reynolds numbers, but because of the secondary effects the correlation is not good.   In scaling up from ¼- to ¾-in. pipe, corrosion rates vary from one-third to twice those predicted on the assumption of constant $D\delta$.   The absolute rates of penetration are small, of the order of $\frac{1}{100}$ in./year, and this degree of approximation might suffice for the purpose of specifying pipe-wall thickness.

In general, where corrosion is diffusion-controlled, in homologous systems at equal Reynolds numbers the average rate of penetration tends to vary inversely as the linear dimension. A model test is then inherently an accelerated corrosion test although materials, concentrations, and temperatures remain the same as in the prototype system. Even at equal fluid velocities, corrosion tends to be somewhat more severe in the smaller pipe.

Concentration Differences

Where the concentration of an electrolyte in contact with a metal varies from place to place, galvanic currents are set up through the electrolyte and in the metal, some areas of the latter becoming anodic and some cathodic. The anodic areas tend to dissolve in the electrolyte unless protected by an adherent and impervious oxide film. This is considered to be the principal mechanism of corrosion of metals by electrolytes.

U. R. Evans (72) has shown that a common source of corrosion due to concentration gradients is differential aeration. A solution exposed to the atmosphere dissolves oxygen, and in an unstirred vessel the concentration tends to be greatest near the surface and decreases with depth. The effect on a metal surface in contact with the solution is that those areas in contact with solution of higher oxygen content become cathodic, those where the oxygen content is low become anodic and corrode. Submerged crevices are particularly subject to corrosion by differential aeration, and the effect is to cause deepening of the crevice and eventual perforation of the metal. In a stirred vessel, the oxygen concentration (except in crevices) is likely to be fairly uniform, but there may be local concentration differences due to the solution of solids or the evaporation of solvent.

If it can be assumed that the current due to concentration differences is so small that anodic or cathodic polarization is negligible, then, other things being equal, the magnitude of the corrosion current will be inversely proportional to the length of the liquid path between the anodic and cathodic areas. That is to say, if in model and prototype vessels the electrolyte concentrations at corresponding points are equal, electrochemical corrosion is likely to be more severe in the model.

Metal-ion Depletion

Müller and Konopicky (74) found that, when a metal is in contact with an electrolyte of uniform concentration which is moving in one region and stagnant in another, an emf is set up between the two regions and corrosion may occur.   This is attributed to the mechanical depletion of metal ions by the fast-moving electrolyte and has been called the *motoelectric effect*.   It has been particularly noted with copper and cuprous alloys, which are relatively immune from corrosion by differential aeration.   With ferrous metals, the effect is likely to be masked by differential aeration.   Inglesent and Storrow (73) attributed to this cause the corrosion of copper evaporator tubes concentrating a glucose liquor, and they measured potential differences of the order of 0.08 volt between the top and bottom of a single tube.

The magnitude of the motoelectric emf, and hence the severity of attack, depends upon differences in velocity or turbulence of the electrolyte adjacent to the metal.   According to Müller and Konopicky, the polarity may vary with the composition of the electrolyte.   Inglesent and Storrow found that their copper evaporator tubes corroded most severely toward the top, where the fluid velocity was greatest.

Erosion

A metal surface in contact with a stream of fluid containing suspended solids is attacked in two ways:

1. Direct abrasion of the metal by the solid particles
2. Continuous removal of an oxide or other protective film so that corrosion by the fluid medium is accelerated

Attack from either cause increases as the fluid velocity is increased.

Suspended fluid particles can also cause erosion, e.g., water droplets in steam or air bubbles in water.   Their effect is primarily due to cause 2.   In the liquid-phase phenomenon known as *cavitation*, low-pressure regions are created behind rapidly moving solid surfaces in which bubbles of low-pressure air or vapor—so-called "vacuum bubbles"—are formed.   These can produce severe erosion of, for example, ships' propellers, their effect being attributed to disturbances of the protective film on the metal.

While it is not yet possible to give quantitative scale relations for the over-all rate of attack by corrosive agents, it will be seen that corrosion due to concentration differences increases as the distance between corresponding electrolyte concentrations is reduced, while, in the case of diffusion control, ion depletion, and erosion, the rate of attack increases with the fluid velocity.   It follows that, if a prototype apparatus and a scale model are exposed to the same corrosive fluid under conditions such that the Reynolds numbers are equal, those corrosion effects which depend on geometry or fluid dynamics will be intensified in the model.   Since the fluid-flow pattern will be the same in both systems, any tendency to localized attack should also be exaggerated in the model.

In the more usual case of a pilot-scale unit operated at much lower Reynolds numbers than the prototype, the fluid-flow pattern will not be the same unless both systems are operating in the region in which the effect of Reynolds number on velocity distribution is negligible— at Reynolds numbers above, say, five times the lower critical value. Where heat- and mass-transfer coefficients in the model and prototype are equal, the average corrosion rates may be expected to be of the same order but regions of localized attack may occur in the prototype which were not foreshadowed in the model.

CHAPTER 19

ANALOGUE MODELS

Certain kinds of analogue models have already been referred to in Chaps. 5 and 12. The present chapter deals with the general theory and application of analogue models, a relatively new and potentially valuable technique as applied to chemical-engineering problems. The mathematical treatment is necessarily more advanced than in the rest of the book and employs the notation of vector analysis.

Ordinary model theory implies a knowledge of the fundamental differential equations sufficient at least for the derivation of scale relations. Whether the systems being compared are homologous* or not, the same kinds of quantities are compared on the small and the large scale. For example, the combustion chamber of a furnace may be simulated by a water model, but the quantities observed in the model and predicted for the prototype are still of the same kind: fluid velocities, flow patterns, pressure drops, etc. There is, however, a further possibility, namely, that the same differential equations may be obeyed by quite different physical or chemical variables. For instance, the differential equations governing irrotational fluid flow are identical in form with those applicable to electrostatic and magnetic fields. It is therefore possible to predict the flow pattern in a large-scale fluid-flow system by plotting lines of force in an electrostatic or magnetic system of similar shape, and this may be a much simpler and quicker procedure than carrying out experiments with a hydrodynamic model. By using a suitable scale factor between corresponding variables and maintaining identical boundary conditions, one may even be able to convert observations of electric or magnetic potential in the electrical system into quantitative predictions of pressure or velocity in the hydrody-

---

* Homologous systems, in the sense intended in this work, are physical systems of which the solid parts are geometrically similar and the physical and chemical properties of the component substances are identical (e.g., same fluids at the same temperatures and pressures). A furnace chamber and a water model represent systems that may be similar but are not homologous.

namic system. The electrical system then constitutes an *analogue model* of the hydrodynamic system.

In general, an analogue model is an experimental apparatus in which certain physical variables are observed and the results used to predict the behavior of variables of a different kind in the prototype system. An analogue model does not necessarily bear any geometrical resemblance to its prototype. It may, for example, consist of a network of electrical resistances, inductances, and capacities designed to obey finite difference equations approximating to the differential equations of the prototype system. As mentioned in Chap. 1, the analogue model in its ultimate development merges into the analogue computer, and at that point the distinction between calculation and experiment vanishes.

The principal kinds of analogue models that have been used up to the present are:

1. Analogues between two-dimensional systems obeying Laplace's or Poisson's equations, e.g., torsion in a bar, displacement of an elastic membrane, steady-state heat conduction, magnetic and electrostatic fields, electric current, viscous flow, and noninertial nonviscous flow

2. Analogues between systems obeying the Fourier equations with time variation and up to three space dimensions, e.g., unsteady-state heat conduction, electric current, and viscous flow of fluids with reservoirs

3. Analogues between mass, momentum, and heat transfer in turbulent-flow systems

4. Analogues between radiation of different wavelengths in geometrically similar systems, and electrical analogues for radiation

5. Analogues for stochastic processes (processes in which the input variable is random in time)

The review of rate equations for chemical reactions given in Chaps. 5 and 7 indicates the possibility of a sixth class, namely, analogues between different chemical reactions obeying similar rate equations. So far as is known, chemical-analogue models have not yet found practical application, although they may do so in the future. For example, chemical-reaction velocities and yields in a dangerously radioactive system might conceivably be investigated by means of an analogous reaction between nonradioactive compounds. Scale factors for chemical analogues are readily deduced from the appropriate dimensionless equations given in Chap. 7 under Chemical Regime [Eqs. (7-24)ff.]. Chemical-analogue models are not treated further in this chapter.

ANALOGUES BETWEEN TWO-DIMENSIONAL SYSTEMS OBEYING LAPLACE'S OR POISSON'S EQUATIONS

Electrostatic Fields

It is a simple consequence of the inverse-square law that

$$\frac{\partial^2 V}{\partial x^2} + \frac{\partial^2 V}{\partial y^2} + \frac{\partial^2 V}{\partial z^2} = -4\pi R \qquad (19\text{-}1a)$$

where V is the electrostatic potential at any point and R is the volume density of electric charge in the neighborhood of the point (x,y,z). In free space where there are no changes, this reduces to Laplace's equation

$$\nabla^2 V = 0 \qquad (19\text{-}1b)$$

if $\nabla^2 =$ the operator $\partial^2/\partial x^2 + \partial^2/\partial y^2 + \partial^2/\partial z^2$. For systems which vary only in the x and y directions, it reduces to

$$\frac{\partial^2 V}{\partial x^2} + \frac{\partial^2 V}{\partial y^2} = 0$$

If there are any conductors bounding the field, these must be equipotentials, i.e., surfaces of $V = $ const. V is related to the electric field $\mathbf{E}$ at any point by the equation

$$\mathbf{E} = -\operatorname{grad} V = -\left(\frac{\partial V}{\partial x}, \frac{\partial V}{\partial y}, \frac{\partial V}{\partial z}\right)$$

If there are dielectrics present of inductive capacity K, Poisson's equation becomes

$$\frac{\partial}{\partial x}\left(K\frac{\partial V}{\partial x}\right) + \frac{\partial}{\partial y}\left(K\frac{\partial V}{\partial y}\right) + \frac{\partial}{\partial z}\left(K\frac{\partial V}{\partial z}\right) = -4\pi R \qquad (19\text{-}2)$$

Across a boundary where the inductive capacity changes, the tangential component of the electric-field intensity $\mathbf{E}$ has the same value $(\partial V_1/\partial S = \partial V_2/\partial S)$, and the normal components of polarization have the same values $(K_1\,\partial V_1/\partial n = K_2\,\partial V_2/\partial n)$.

Magnetic Fields

Similarly for magnetostatic systems in a region which contains no magnetic matter

$$\nabla^2 \Omega = 0$$

where Ω is the magnetic potential, which is related to the magnetic field by the equation*

$$\mathbf{H} = -\operatorname{grad}\,\Omega = -\left(\frac{\partial\Omega}{\partial x},\frac{\partial\Omega}{\partial y},\frac{\partial\Omega}{\partial z}\right)$$

If there is magnetic material of permeability M, the magnetic induction I is related to Ω by the equation

$$\mathbf{I} = -M\operatorname{grad}\,\Omega$$

and Poisson's equation has the form

$$\frac{\partial}{\partial x}\left(M\frac{\partial\Omega}{\partial x}\right) + \frac{\partial}{\partial y}\left(M\frac{\partial\Omega}{\partial y}\right) + \frac{\partial}{\partial z}\left(M\frac{\partial\Omega}{\partial z}\right) = 0 \qquad (19\text{-}3)$$

Electric Current

In the steady flow of electric current through a three-dimensional conductor of specific resistance Γ, Ohm's law gives the equation

$$\mathbf{i} = \frac{1}{\Gamma}\operatorname{grad}\,e = \frac{1}{\Gamma}\left(\frac{\partial V}{\partial x},\frac{\partial V}{\partial y},\frac{\partial V}{\partial z}\right) \qquad (19\text{-}4a)$$

where V is the electric potential and $\mathbf{i}$ is the current vector. The conservation of electric current gives the equation

$$\operatorname{div}\mathbf{i} = 0 \qquad \text{that is} \qquad \frac{\partial i_x}{\partial x} + \frac{\partial i_y}{\partial y} + \frac{\partial i_z}{\partial z} = 0 \qquad (19\text{-}4b)$$

which gives, from (19-4a),

$$\frac{\partial}{\partial x}\left(\frac{1}{\Gamma}\frac{\partial V}{\partial x}\right) + \frac{\partial}{\partial y}\left(\frac{1}{\Gamma}\frac{\partial V}{\partial y}\right) + \frac{\partial}{\partial z}\left(\frac{1}{\Gamma}\frac{\partial V}{\partial z}\right) = 0 \qquad (19\text{-}4c)$$

So if Γ is constant through the space region, this again gives V obeying Laplace's equation,

$$\nabla^2 V = 0$$

If there is a boundary having a very much greater conductivity than the main material this will be an equipotential, i.e., a surface $V = \text{const.}$

Displacement of an Elastic Membrane or Soap Film

If an elastic membrane or a soap film stretched by a uniform tension τ per unit length is fixed with its outer boundary in one plane and

---

* J. H. Jeans, "Mathematical Theory of Electricity and Magnetism," 5th ed., p. 371, Cambridge University Press, New York, 1925.

possibly with an inner boundary at another plane and the air acts on one side of it with a pressure p above that on the other side, the membrane will become a curved surface. Let the displacement from the fixed plane of any point on the membrane whose coordinates in the fixed plane are (x,y) be z, and suppose z is small compared with x and y. Then the equation expressing the static equilibrium of an element of area $dx\,dy$ is

$$p\,dx\,dy = \tau\,dy\left(\frac{\partial z}{\partial x} + \frac{\partial^2 z}{\partial x^2}\,dx - \frac{\partial z}{\partial x}\right) + \tau\,dx\left(\frac{\partial z}{\partial y} + \frac{\partial^2 z}{\partial y^2}\,dy - \frac{\partial z}{\partial y}\right)$$

(19-4d)

so

$$\frac{p}{\tau} = \frac{\partial^2 z}{\partial x^2} + \frac{\partial^2 z}{\partial y^2}$$

If the pressure difference p is zero, then this reduces to Laplace's equation governing the small displacement z.

$$\frac{\partial^2 z}{\partial x^2} + \frac{\partial^2 z}{\partial y^2} = 0 \qquad (19\text{-}4e)$$

Hydrodynamic Systems

In the case of fluid flow, the basic differential equations governing the motion are the Navier-Stokes equations discussed in Chap. 5:

$$\rho\left(\frac{\partial u}{\partial t} + u\frac{\partial u}{\partial x} + v\frac{\partial u}{\partial y} + w\frac{\partial u}{\partial z}\right) = X\rho - \frac{\partial p}{\partial x} + \mu\nabla^2 u$$
$$+ \mu\frac{\partial}{\partial x}\left(\frac{\partial u}{\partial x} + \frac{\partial v}{\partial y} + \frac{\partial w}{\partial z}\right) \quad (19\text{-}5)$$

where X is the x component of the external-force system (e.g., in the case where gravity provides the external-force system, $X = g\cos\alpha$, where α is the angle between the x axis and the direction of gravity), with similar equations for the y and z axes, and the equation of continuity

$$\frac{1}{\rho}\frac{d\rho}{dt} + \frac{\partial u}{\partial x} + \frac{\partial v}{\partial y} + \frac{\partial w}{\partial z} = 0 \qquad (19\text{-}6)$$

For a nonviscous incompressible fluid, these equations reduce to

$$\rho\left(\frac{\partial u}{\partial t} + u\frac{\partial u}{\partial x} + v\frac{\partial u}{\partial y} + w\frac{\partial u}{\partial z}\right) = X\rho - \frac{\partial p}{\partial x} \qquad (19\text{-}7)$$

$$\frac{\partial u}{\partial x} + \frac{\partial v}{\partial y} + \frac{\partial w}{\partial z} = 0 \qquad (19\text{-}8)$$

If we assume that the impressed force is due to a potential V', so that the force acting at a point is the negative gradient of V',

$$X = -\frac{\partial V'}{\partial x} \qquad Y = -\frac{\partial V'}{\partial y} \qquad Z = -\frac{\partial V'}{\partial z} \qquad (19\text{-}9)$$

To integrate the equations, we can introduce a space function

$$\boldsymbol{\omega} \equiv (\xi, \eta, \zeta)$$

defined by the equations

$$\frac{\partial w}{\partial y} - \frac{\partial v}{\partial z} = 2\xi \qquad \frac{\partial u}{\partial z} - \frac{\partial w}{\partial x} = 2\eta \qquad \frac{\partial v}{\partial x} - \frac{\partial u}{\partial y} = 2\zeta \quad (19\text{-}10)$$

Then Eq. 19-7 becomes

$$\frac{\partial u}{\partial t} - 2v\zeta + 2w\eta + \frac{\partial H}{\partial x} = 0 \qquad (19\text{-}11a)$$

$$\frac{\partial v}{\partial t} - 2w\xi + 2u\zeta + \frac{\partial H}{\partial y} = 0 \qquad (19\text{-}11b)$$

$$\frac{\partial w}{\partial t} - 2u\eta + 2v\xi + \frac{\partial H}{\partial z} = 0 \qquad (19\text{-}11c)$$

where

$$H = \int \frac{d\rho}{\rho} + V' + \frac{1}{2}(u^2 + v^2 + w^2) \qquad (19\text{-}12)$$

H is thus the sum of the pressure head, the potential head, and the velocity head.

If we eliminate H between (19-11b) and (19-11c) by partial differentiation with respect to z and y respectively, we obtain

$$\frac{\partial \xi}{\partial t} - \xi \frac{\partial u}{\partial x} - \eta \frac{\partial v}{\partial x} - \zeta \frac{\partial w}{\partial x} + \xi \left(\frac{\partial u}{\partial x} + \frac{\partial v}{\partial y} + \frac{\partial w}{\partial z} \right) = 0$$

The last term vanishes because of the equation of continuity (19-8) so that

$$\frac{\partial \xi}{\partial t} - \xi \frac{\partial u}{\partial x} - \eta \frac{\partial v}{\partial x} - \zeta \frac{\partial w}{\partial x} = 0$$

with two similar equations for the y and z axes.

These equations mean that, if the function $\boldsymbol{\omega}$ vanishes

$$(\omega^2 = \xi^2 + \eta^2 + \zeta^2 = 0)$$

throughout the fluid at one time, then it will always be zero and the fluid motion is called *irrotational*. The vector $\boldsymbol{\omega}$ is called the *vorticity*. If we confine ourselves to a motion initially made irrotational,

$$\frac{\partial w}{\partial y} = \frac{\partial v}{\partial z} \qquad \frac{\partial u}{\partial z} = \frac{\partial w}{\partial x} \qquad \frac{\partial v}{\partial x} = \frac{\partial u}{\partial y} \qquad \text{by (19-10)}$$

In this case, a function ϕ can be found such that

$$\frac{\partial \phi}{\partial x} = u \qquad \frac{\partial \phi}{\partial y} = v \qquad \frac{\partial \phi}{\partial z} = w \qquad (19\text{-}13)$$

(or vector velocity $\mathbf{u} = \text{grad } \phi$), since such a function must satisfy $\partial w/\partial y = \partial v/\partial z$, etc., because $\partial w/\partial y = \partial^2 \phi/\partial y\,\partial z$, $\partial v/\partial z = \partial^2 \phi/\partial z\,\partial y$.

ϕ is called the *velocity function;* it bears the same relation to the velocity vector as the electrostatic potential bears to the electric field. Surfaces of equal ϕ are surfaces everywhere at right angles to the direction of flow.

Then, from the equation of continuity (19-8),

$$\nabla^2 \phi = 0 \qquad (19\text{-}14)$$

The existence of the function ϕ is thus an integration of the dynamic equations of motion and its obedience to Eq. (19-14) is a consequence of the conservation of fluid. If, as well as being irrotational, the motion is steady, i.e., the velocity at any point in space does not change with time, $\partial u/\partial t = 0$ and so, from Eqs. (19-1b) and (19-11a) to (19-11c), H is constant everywhere throughout space. This is Bernoulli's theorem of energy conservation.

If now we consider two-dimensional motion for which the conservation equation is

$$\frac{\partial u}{\partial x} + \frac{\partial v}{\partial y} = 0 \qquad (19\text{-}8a)$$

we can define a second function ψ of (xy) such that

$$u = \frac{\partial \psi}{\partial y} \qquad v = \frac{\partial \psi}{\partial x} \qquad (19\text{-}15)$$

since the existence of such a function makes (u,v) satisfy (19-8b). ψ is called the *stream*, or *current, function*. Its physical interpretation is that $\psi_a - \psi_b$ equals the volumetric flow rate in cubic feet per hour across any curve $y = f(x)$ between the points A and B. This flow rate is necessarily the same whatever curved surface is taken because of continuity. Hence, lines of equal ψ are streamlines since no fluid flows across them. They correspond to the lines of force in the electrostatic or magnetostatic fields.

From Eq. (19-10c), for the irrotational motion $\zeta = 0$,

$$\frac{\partial v}{\partial x} = \frac{\partial u}{\partial y}$$

and hence, from (19-15),

$$\nabla^2 \psi = 0$$

From Eqs. (19-13) and (19-15),

$$u = \frac{\partial \phi}{\partial x} = - \frac{\partial \psi}{\partial y} \qquad v = \frac{\partial \phi}{\partial y} = - \frac{\partial \psi}{\partial x}$$

so that ϕ and ψ are conjugate functions of (x,y); the curves ϕ = const and ψ = const form an orthogonal system. The interchange of ϕ and ψ describes another possible state of plane flow in which the velocity at every point is turned through a right angle but remains equal in magnitude.

The analogy between nonviscous irrotational flow in two dimensions and electrostatic or magnetostatic fields can thus be used by equating either of the functions ψ or ϕ to the electric or magnetic potential and using the other function to describe the lines of force, but inserting the appropriate boundary conditions in either case.

In the hydrodynamic case, we can only have a boundary obstructing the flow which is necessarily a curve of constant ψ. On the other hand, in the electrostatic case there can be changes in dielectric constant for which there is no hydrodynamic analogy and conductors which are equipotentials (V = const). Hence, the analogy between electrostatic fields and irrotational nonviscous fluid flow is necessarily confined to two dimensions for which the conjugate functions ψ, ϕ are available, and necessarily reverses the analogy; i.e., the analogy is as follows:

Flow systems	Electrostatic systems
Streamlines	Equipotentials
Flow surfaces	Lines of force
Impassable barrier	Conductor

The analogy in this form has been used by Relf (113) to investigate the ideal hydrodynamic flow round an obstacle such as an airfoil by means of an electrostatic analogue model. To explore the equipotentials, he did not use a perfectly nonconducting dielectric between the parallel walls and the metal obstacle but used a substance of low conductivity—stationary tap water in a concrete tank. He used an alternating field of audible frequency (200 cycles/sec) and balanced the potential on the exploring electrode against that on a potentiometer wire joining the side plates by means of an amplifier and phone indicator. He could locate the streamlines (equipotentials) to $\frac{1}{100}$ in. in a tank 2 ft 6 in. wide in this way. He was also able to investigate the case of circulation flow around the airfoil without any transverse flow by connecting both the side plates to one side of the a-c supply and the model to the other side.

Langmuir (159) has also developed the electrolytic tank as a means of plotting equipotentials satisfying Laplace's equation in two dimensions. A saturated solution of copper sulfate with a little sulfuric acid is used and a 400-cycle alternating current of density about 5 ma/cm^2 is passed between the electrodes.

Hele-Shaw (138) developed empirically a converse analogue in which a slow-flowing viscous fluid is used to give a solution to the equations of two-dimensional magnetostatic fields in the presence of materials of different permeability or of electrostatic fields in the presence of materials of different dielectric constants. He based the analogy on the fact that it gave the correct streamlines for a case where the boundary conditions are sufficiently simple to enable an exact mathematical solution of Laplace's equation to be obtained, that of an infinite cylinder of elliptic section of permeability 100 placed in an originally uniform magnetic field with its major axis along the field. He showed that, when there is parallel flow along the x axis of a viscous fluid between flat plates very close together, the volume q flowing through per second per unit width of a layer of thickness d is given by

$$q = \frac{d^3}{12\mu} \frac{dp}{dx}$$

and he concluded that, if the cube of the distance between the plates were made proportional to the permeability as a function of x and y, then the streamlines would coincide with the magnetic lines of force. The fluid was introduced in a uniform stream across one end of the rectangular model and removed across the other, corresponding to the uniform magnetic field remote from the disturbing objects. The flow pattern was made visible by introducing alternate streaks of transparent and colored fluid at the end of the model.

Hele-Shaw did not attempt to show that the equation governing the motion must be the same as that governing the magnetic field, but Stokes (142) pointed out that this is in fact a consequence of the low velocities used, which cause the terms depending on the square of the velocity in Eq. (19-5) to be negligible. A simple proof of this can be derived as follows: If we have an irrotational steady-state flow in the absence of an external force system of an incompressible fluid, then (19-5) reduces to

$$\frac{\partial p}{\partial x} = \mu \nabla^2 u$$

$$\frac{\partial p}{\partial y} = \mu \nabla^2 v \tag{19-16}$$

$$\frac{\partial p}{\partial z} = \mu \nabla^2 w$$

and we have the conservation equation $\partial u/\partial x + \partial v/\partial y + \partial w/\partial z = 0$. Now let z be the direction perpendicular to the plates bounding the flow, and assume that the velocity gradients in the z direction are much greater than those in the x and y directions; i.e., the flow is determined by viscous resistance from the walls and not between lateral layers of liquid. Then the equations of motion become

$$\frac{\partial p}{\partial x} = \mu \frac{\partial^2 u}{\partial z^2}$$

$$\frac{\partial p}{\partial y} = \mu \frac{\partial^2 v}{\partial z^2} \qquad (19\text{-}17)$$

The distance apart of the plates d is a function of x and y, but we assume it does not vary sharply in a distance comparable with d except at a sharp boundary between two materials of different permeability. Then we can integrate the equations of motion twice with respect to z, inserting the boundary conditions $\partial u/\partial z$, $\partial u/\partial z = 0$ when $z = 0$ (central plane) and $u, v = 0$ when $z = \pm \frac{1}{2}d$ (walls), to obtain

$$\mu U_m = \frac{\partial p}{\partial x} \frac{d^2}{8} \qquad (19\text{-}18a)$$

$$\mu U = \frac{\partial p}{\partial x}\left(\frac{d^2}{8} - \frac{z^2}{2}\right) \qquad (19\text{-}18b)$$

$$\mu V_m = \frac{\partial p}{\partial y} \frac{d^2}{8} \qquad (19\text{-}18c)$$

$$\mu V = \frac{\partial p}{\partial x}\left(\frac{d^2}{8} - \frac{z^2}{2}\right) \qquad (19\text{-}18d)$$

where U_m, V_m are the flow velocities halfway between the bounding planes.

The quantity of fluid flowing in the direction of the x axis across a small element $d\,\delta y$ is obtained by integrating Eq. (19-18b) across the thickness d, as

$$q_x\,\delta y = \frac{d^3}{12\mu} \frac{\partial p}{\partial x} \delta y$$

Hence, the equation for conservation of fluid is

$$\frac{\partial q_x}{\partial x} + \frac{\partial q_y}{\partial y} = 0$$

or

$$\frac{\partial}{\partial x}\left(d^3 \frac{\partial p}{\partial x}\right) + \frac{\partial}{\partial y}\left(d^3 \frac{\partial p}{\partial y}\right) = 0 \qquad (19\text{-}19)$$

This equation in p (the static pressure) is identical in form with Eq. (19-2) governing the electric potential V when $p = 0$ (no free charges) and with Eq. (19-3) governing the magnetic potential Ω provided d^3 is made the same function of (x,y) as K or M, respectively. Lines of equal pressure p in the Hele-Shaw model are thus equipotentials in the magnetic or electrostatic field, and the flow lines, which always flow along the maximum pressure gradient, coincide with the lines of force.

Heat Flow

The conduction of heat in solids obeys the Fourier equation

$$c\rho \frac{\partial T}{\partial t} = \frac{\partial}{\partial x}\left(k \frac{\partial T}{\partial x} \right) + \frac{\partial}{\partial y}\left(k \frac{\partial T}{\partial y} \right) + \frac{\partial}{\partial z}\left(k \frac{\partial T}{\partial z} \right) \qquad (19\text{-}20)$$

where T = temperature at time t at point P (x,y,z)
c = specific heat per unit mass at P
ρ = density at P
k = thermal conductivity of solid at P
This equation is simply the expression for the conservation of heat, and the flow of heat $\mathbf{q}$ is given from the temperature distribution by the equation $\mathbf{q} = k$ grad T. The condition across a boundary between two solids is

$$k_1 \frac{\partial T}{\partial n} = k_2 \frac{\partial T_2}{\partial n} \qquad (19\text{-}21)$$

In steady-state heat flow, the left-hand side of (19-20) vanishes, and (19-20) becomes identical with (19-2) for the case of $p = 0$ and with (19-3), with temperature appearing as electric or magnetic potential, thermal conductivity as dielectric constant or magnetic permeability, and lines of heat flow as lines of force. A perfectly insulated surface is one for which $\partial T/\partial n = 0$; it must thus coincide with a flow line.

When the thermal conductivity is independent of position and temperature, (19-20) reduces to

$$\nabla^2 T = 0$$

which is Laplace's equation once again.

Hence, models giving a solution of Laplace's equation with the appropriate boundary conditions can be used to find the flow of heat in steady-state conduction through awkward shapes provided the thermal conductivity is constant. The analogue between heat flow and electrostatic fields has been used by Langmuir, Adams, and Meikle (159) to solve the problems of two-dimensional heat flow in a thick corner. The isothermal source and sink of heat on the inside and outside of the corner were represented by the metallic electrodes, the heat flow lines across the ends of the corner and the base were represented by nonconducting sides and base of the tank, and an electrolyte was used with alternating current and an exploring probe to find the equipotentials.

Kayan (157) used the analogy between steady-state heat conduction and steady-state electric-current flow in a two-dimension resistor. He wished to determine the isotherms in a building structure at a corner between four rooms, two at one temperature and two at another, when the corner was built of layers of concrete and layers of insulating material. He used a sheet of metallic foil as the resistor and reduced the specific resistance of the part corresponding to the insulating material by cutting out a regular mesh of square holes in it, leaving narrow webs between the holes. The mesh size was successively increased for a fixed web thickness by cutting out the inside webs of each large square. The resistance of each mesh size was separately determined. This enabled the isotherms corresponding to the true value of the thermal conductivity of the insulating layer to be determined by interpolation. The effect of the thermal resistance of the air boundary layer in the rooms was included by introducing a layer of solid foil the thickness of which, relative to the thickness of the solid foil representing the concrete, had the same ratio as the inverse ratio of their thermal conductances in chu/(hr)(ft^2)($°$C). It worked out that the air resistance was equivalent to 3 in. of concrete; so if the model had the same linear scale as the actual corner, the solid sheet representing the air resistance was 3 in. wide. Beyond the air resistance were strongly conducting electrodes representing the hot and cold rooms which were connected to a battery source of direct current. The equipotentials (isothermals) were explored with a movable probe connected via a galvonometer to a potentiometer slide wire across the battery.

Jakob (153) used a magnetic analogue for heat flow in which an electromagnet had a core of sheets of shape imitating the surface of the conducting object. Iron filings on a cardboard sheet between the poles gave the magnetic lines of force, which were equivalent to the lines of heat flow. Isotherms could be determined by constructing the orthogonal set of lines.

Wilson and Miles (175) have used the membrane analogy for the solution of two-dimensional heat-conduction problems. They used a soap film without any pressure difference across it so that Laplace's equation [Eq. (19-4)] governs the displacement z. The inner and outer edges of the annular film were at different heights, corresponding to the temperature difference across the heat-conducting hollow cylinder. The height of the film between could be explored by means of a micrometer screw with a point working in a flat glass plate movable across the model and a device for marking the position of the screw on a sheet of paper of the same size as the model. The method can be used to give the isotherms in complex shapes quite accurately and speedily but is obviously unable to deal with problems involving a combination of materials of different thicknesses.

Shearing Stress in a Twisted Bar [Elastic and Plastic (268)]

Consider first an elastic bar with any cross section $y = f(x)$ but with this cross section constant along its length, and suppose this bar is given a uniform twist so that the cross section $z = $ const rotates about the z axis—in unit length a particular section rotates by an angle θ. Thus, the displacement of a point P (xyz) is given by

$$\xi = -\theta yz \qquad \eta = \theta xz \qquad \zeta = \theta \phi(xy)$$

where $\phi(xy)$ is the distortion of the originally plane surface $z = $ const. Hence, the shearing strain per unit length is

$$\gamma_{xz} = \frac{\partial \xi}{\partial z} + \frac{\partial \zeta}{\partial x} = \theta \left(\frac{\partial \phi}{\partial x} - y \right)$$

$$\gamma_{yz} = \frac{\partial \eta}{\partial z} + \frac{\partial \zeta}{\partial y} = \theta \left(\frac{\partial \phi}{\partial y} + x \right)$$

Hence, by Hooke's law, the shearing stresses are given by

$$S_x = G\gamma_{xz} = G\theta \left(\frac{\partial \phi}{\partial x} - y \right)$$

$$S_y = G\theta \left(\frac{\partial \phi}{\partial y} + x \right)$$

where G is the rigidity modulus. Thus, eliminating ϕ, we obtain

$$\frac{\partial S_x}{\partial y} - \frac{\partial S_y}{\partial x} = -2G\theta \tag{19-22}$$

which is independent of x and y.

Also, the equation of equilibrium of the stresses acting on an element $dx\, dy$ is

$$\frac{\partial S_x}{\partial x} + \frac{\partial S_y}{\partial y} = 0 \tag{19-23}$$

Equations (19-22) and (19-23) are exactly analogous to the hydrodynamic equations (19-8b) and (19-15), with S_x and S_y replacing u and v. We can thus introduce a function Σ defined by the equations

$$S_x = \frac{\partial \Sigma}{\partial y} \qquad S_y = -\frac{\partial \Sigma}{\partial x} \tag{19-24}$$

which automatically satisfies (19-23) and when inserted into (19-22) gives

$$\frac{\partial^2 \Sigma}{\partial y^2} + \frac{\partial^2 \Sigma}{\partial x^2} = -2G\theta \tag{19-25}$$

The boundary conditions for this function are that the shearing stress at the edge $y = f(x)$ shall act along the edge; that is, $S_y/S_x = dy/dx$ when $y = f(x)$, which means that $\Sigma = $ const along the boundary.

The simplest analogy for this case is the elastic membrane discussed earlier. If a pressure p is applied to the membrane so that p/T is set equal to $-2G\theta$, then the observed value of the displacement z of a membrane bounded by a curve equal to the outline $y = f(x)$ of the twisted bar will give the function Σ, and the shear stresses S_x and S_y can be derived by differentiation as in Eq. (19-24). Along a line of constant $\Sigma, (\partial\Sigma/\partial y)\, dy + (\partial\Sigma/\partial x)\, dx = 0$; so

$$\frac{dy}{dx} = \frac{-\partial\Sigma/\partial x}{\partial\Sigma/\partial y} = \frac{S_y}{S_x} \qquad \text{from (19-24)}$$

Thus, the shearing stress acts along the tangent to the line of constant Σ and its magnitude $\sqrt{S_x{}^2 + S_y{}^2} = \sqrt{(\partial\Sigma/\partial y)^2 + (\partial\Sigma/\partial x)^2}$ is equal to the largest slope of Σ which is in the direction at right angles to the line of constant Σ.

This method has been used by Griffith and Taylor (263 to 267) to determine the torsional stiffness and strength of cylindrical bars, flexure of beams, and torsion and flexure of hollow shafts. Neubauer and Boston (269) have used it for twist-drill sections.

Now consider the case where part of the bar is twisted beyond the elastic limit. We have to assume that the condition of plastic yield is that the shear stress reaches a certain magnitude k and then remains constant, while the material yields indefinitely. The condition of plastic yield is then

$$\sqrt{S_x{}^2 + S_y{}^2} = k \tag{19-26}$$

while for values of S_x, S_y below this point the material remains elastic and obeys Hooke's law.

The condition of static equilibrium (19-23) must still hold in the plastic region so that we can still define the function Σ as in (19-24), but the differential equation that must be satisfied by Σ is no longer (19-25) but is now given by (19-26) to be

$$\left(\frac{\partial\Sigma}{\partial x}\right)^2 + \left(\frac{\partial\Sigma}{\partial y}\right)^2 = k^2$$

This is simply the condition that

$$\frac{\partial \Sigma}{\partial n} = k$$

where n is the direction of maximum change of Σ. In terms of the membrane analogy, the maximum slope of the membrane is constant throughout the plastic region. The plastic stress surface is thus simply a surface of constant maximum slope bounded by the boundary of the bar. It is the surface of a sand heap (270) piled on the cross section of the bar to have everywhere the maximum angle of repose.

If a bar is twisted so that part of it is deformed elastically and part plastically, the shearing stress must be equal to k on both sides of the boundary between the two types of deformation. Hence, the membrane analogy can be used to determine the boundary between the zones if a roof corresponding to the fully plastic shape is placed over the membrane and the pressure corresponding to various degrees of twist applied to the membrane. The parts of the membrane in contact with the roof give the plastically distorted region, and the parts where the tension in the membrane holds it away from the roof give the elastically distorted region. Nádai and Friedman (268) have used this technique for square, rectangular, triangular, elliptical, and circular bars with circular grooves, and Coker (262) has used it for twisted shafts with keyways.

ANALOGUES BETWEEN SYSTEMS OBEYING THE FOURIER EQUATION WITH TIME VARIATION

Electric Analogue

The electric analogue for heat conduction has been carried to the point of development where it gives a thoroughly accurate and convenient method of solving practical industrial problems such as the freezing of a metal ingot. Paschkis (169) and Beuken (147) have been largely responsible for this development. The Fourier equation (19-20) for unsteady-state heat transfer when k, c, and ρ are independent of temperature reduces to $\partial T/\partial t = a\nabla^2 T$, where $a = k/c\rho$, the thermal diffusivity. The equation governing the emf V in a single long conductor with distributed resistance R_e and distributed capacity C_e to earth all along it is

$$\frac{\partial V}{\partial t} = \frac{1}{R_e C_e} \frac{\partial^2 V}{\partial x^2}$$

The analogy is thus between the two following:

Electric	Thermal
Voltage V	Temperature T
Current i	Heat flow q
Electric capacity C_e	Thermal capacity $C_t = C\rho$
Charge	Heat
Ohm's law $i = \dfrac{\Delta V}{R_e}$	Conduction law $q = \dfrac{\Delta T}{1/k}$

There is nothing in the thermal system corresponding to inductance, which gives a term in $\partial^2 V / \partial t\, y$ in the electric system and allows the possibility of oscillation; so the equivalent electric circuit must not have any inductance. To use an analogy with a fully distributed resistance and capacity would, however, make the separate adjustment of R_e and C_e difficult and also limit the system to a one-dimensional flow; so the next step is to replace the conductor by a set of resistances with a condenser connected to earth from the center of each one. This amounts to replacing the differential equation by a stepwise difference equation, physically analogous to the method of calculation by finite-difference steps.* The smaller and more numerous the steps, the more accurate the approximation.

A two- or three-dimensional problem can now be solved by connecting either a flat net of resistances with capacities to earth from each mesh point or a series of such nets with corresponding points in each net connected together by further resistances. It is even possible (158) to connect together a series of three-dimensional structures to form a four-dimensional one and represent time as a fourth space coordinate. In this case, the model is operated with steady alternating current, but this step, although mathematically interesting, has not been so fruitful as the Paschkis one. Paschkis can choose the time

---

* Mathematically, the approximation implies the assumption that, if $n - 1$, n, $n + 1$ are three points at equal intervals along the ordinate x of a curve $V = f(x)$, then

$$\frac{V_{n+1} + V_{n-1} - 2V_n}{(\Delta x)^2}$$

is an approximation to $\partial^2 V / \partial x^2$ since it represents the extent to which the mean of $V_n + 1$ and $V_n - 1$ differs from V_n. Thus, if $C_e \, \Delta x$ and $R_e \, \Delta x$ are replaced by a finite capacity $= C_e{}^1$ and $R_e{}^1$, an exactly analogous approximation is obtained with a series of steps.

scale in the electric analyses arbitrarily compared with that in the thermal systems; thus, let the scales be chosen as follows:

Thermal system		Electric analogue
$\Delta T = 1°C$ or $1°F$	$\equiv$	$\Delta V = $ **V** volts*
$q = 1$ chu/(ft²)(hr)	$\equiv$	$i = $ **i** amp
$C_t = C\rho = 1$ chu/(ft³)(°C)	$\equiv$	$C'_e = $ **C** farads
$R_t = \dfrac{1}{k} = 1$ ft²-°C-hr/chu	$\equiv$	$R'_e = $ **Γ** ohms

* Here boldface type does not, as elsewhere in the book, denote prototype/model ratios, but indicates the number of electrical units corresponding to one thermal unit.

Then, if the heat conduction law $q = \Delta T/R_t$ is to be equivalent to Ohm's law, we must have

$$i = \frac{V}{\Gamma} \qquad (19\text{-}27)$$

while if the capacity equations $C'_e = \int i \, dt/\Delta V$ and $C_t = \int q \, dt/\Delta T$ are to be equivalent (where t is in hours in the heat equation and seconds in the electric one), we must have

$$C = \frac{it}{V} = \frac{t}{\Gamma} \qquad \text{by (19-27)} \qquad (19\text{-}28)$$

Thus, the scale factors for three of the electric quantities, say, **v**, **c**, and **Γ**, can be freely chosen; then the fourth is fixed by (19-27), and the time scale in the model is fixed by (19-28). Conversely, two scale factors and the time scale **t** can be freely chosen for convenience, and then the other two electric scale factors are fixed by the two equations.

Either the boundary condition for the heat transfer can be automatically inserted by using an appropriate resistance for the boundary thermal resistance from a region suddenly raised in temperature, or any required curve

$$T_s = f(t)$$

can be duplicated by hand manipulation of a potentiometer since the time scale can be made appropriate for such manipulation.

Paschkis checked the analogy against temperature measurements on a cork-insulated pipe into which steam was suddenly introduced and then constructed a permanent electric analogue at Columbia University with 15 groups of condensers, each group giving capacities from 0.1 to 20 μf in steps of 0.1 μf, and 15 groups of resistors giving 100 to 1,111,000 ohms in steps of 100 ohms. These groups can be connected up to give a 15-step solution to a large range of problems when direct current variable from 130 volts to 0 is fed in and the voltages are recorded by amplifiers taking less than 10^{-9} amp from the circuit. The total heat input is observed by measuring total charge in coulombs on another amplifier, which actually integrates voltage-time. Thus Paschkis (168) has used the method extensively for the evaluation of the temperatures and movement of the solidification front in the freezing of steel and nonferrous ingots. A two-dimensional solution for a square ingot in a square mold but without end effects is made, and the ingot half thickness is divided into four steps so that there are 16 meshes in the quarter of the square and another 20 in the corner of the mold. Each element of the steel is arranged initially to have resistance and capacity corresponding to the properties of the liquid steel, but when the voltage drop at the mesh point is observed to be that corresponding to the liquidus temperature, an extra capacity corresponding to the heat of fusion spread over the liquidus-solidus range is switched in by hand. Halfway through fusion, the effective resistance is changed to that corresponding to the thermal conductivity of the solid metal, and at the completion of fusion, i.e., when the solidus temperature is reached, the fusion heat capacity is switched out and the capacity corresponding to the solid steel inserted. A resistance corresponding to the formation of an air gap between the casting and the mold can be introduced at the appropriate time.

Viscous Fluid Flow in Fine Tubes

The analogy between heat flow and viscous noninertial fluid flow has also been used for the rapid evaluation of unsteady-state heat-flow problems. Again the differential equation is replaced by a stepwise approximation, and the resistances are capillary tubes of various lengths which obey Poiseuille's law, flow $\propto$ head, while the capacities are vertical fluid reservoirs. Moore's Hydrocal (162) used water as the flowing fluid with capillaries of about 0.05 in. diameter as the resistances and had a device by which the whole flow process could be stopped ("frozen") after any desired time increment and the resistances and stand-pipe capacities appropriately readjusted to allow for changes in thermal conductivity or heat capacity with temperature. The capacity changes could be made by tilting the standpipes at different angles.

Coyle (149) has developed a similar analogue in which it is the viscous flow of air in fine capillaries which simulates heat flow. The air is displaced from reservoirs by light oil, and the head of the oil in the reservoir below that in an open constant-head tank represents the temperature difference between the given element and a fixed datum temperature. The purpose in this case is to develop an analogue which will automatically handle variations of specific heat and thermal conductivity of the conductor with temperature. Variations in specific heat C with T can clearly be handled directly by making the reservoir cross-sectional area vary appropriately with height. This follows from Eq. (19-20), namely,

$$c\rho \frac{\partial T}{\partial t} = \frac{\partial}{\partial x}\left(k\frac{\partial T}{\partial x}\right) + \frac{\partial}{\partial y}\left(k\frac{\partial T}{\partial y}\right) + \frac{\partial}{\partial z}\left(k\frac{\partial T}{\partial z}\right)$$

When $c\rho$ is a function of T, the solution will be exactly analogous to making the reservoir change in capacity a function of height. To deal with the case $k = f(T)$, however, it is necessary to introduce an intermediate variable

$$T^1 = \int_0^T \frac{k}{k_0}\,dT \tag{19-29}$$

which reduces Eq. (19-20) to the form

$$\frac{c\rho}{k_0}\frac{\partial T^1}{\partial t} = \frac{\partial^2 T^1}{\partial x^2} + \frac{\partial^2 T^1}{\partial y^2} + \frac{\partial^2 T^1}{\partial z^2} \tag{19-30}$$

Thus, the equation in T^1 is one with a variable diffusivity $k/\rho c = f(T^1)$, and this can be solved by making the reservoir areas the appropriate functions of height. The analogue then gives the solution to Eq. (19-30), and the corresponding temperatures can be determined from Eq. (19-29).

Coyle makes the reservoirs of rectangular slots in a $\frac{1}{2}$-in. brass plate faced front and back with Perspex and adjusts the cross-sectional area with height by inserting a suitably cast plaster-of-paris block.

The electric analogue has the clear advantage over the fluid-flow ones that accurately adjusted resistances and capacities can be obtained ready-made; but the possibility of building the effects of changes of diffusivity with temperature into the model which Coyle's technique offers is clearly of great potential advantage, especially if one has to solve a large number of problems for the same material.

ANALOGUES BETWEEN HEAT, MATTER, AND MOMENTUM TRANSFER

Two cases must be distinguished, those in which the heat and matter transfers are to a boundary wall, and those in which fluids are mixing and exchanging heat in space essentially away from solids. In the former case, there is in general a turbulent core to the flow, a transition layer, and a laminar layer on the surface of the solid. In the laminar layer, the basic equations for mass and momentum and heat transfer are

$$N_A = -D \frac{dc}{dy}$$

$$S = -\mu \frac{dU}{dy} = -\nu \frac{d(\rho U)}{dy} \qquad (19\text{-}31)$$

$$q = -k \frac{dT}{dy} = -\frac{k}{C_p \rho} \frac{d(C_p \rho T)}{dy}$$

where N_A = mass-transfer rate, lb moles/hr/ft²

S = momentum-transfer rate or shear stress, poundals/ft²

q = heat-transfer rate, chu/(hr)(ft²)

c = concentration of diffusing material, lb moles/ft³

U = velocity of stream parallel to surface at distance y from surface, ft/hr

T = temperature of stream at distance y from surface, °C

C_p = specific heat per unit volume at constant pressure

D, μ, ν, k = diffusion coefficient, absolute and kinematic viscosities, and thermal conductivity of stream fluid

Similarly, in the full turbulent region where matter, heat, and momentum are being transferred by eddies, the equations are

$$N_A = -E \frac{dc}{dy}$$

$$S = -\varepsilon \frac{d(\rho U)}{dy} \qquad (19\text{-}32)$$

$$q = -E_H C_p \rho \frac{dT}{dy}$$

where E, ε, and E_H are the eddy diffusivities for matter, momentum, and heat, respectively (all with dimensions in square feet per hour). Now, if one were to assume that in a turbulent-gas stream these three properties are carried by eddies which move an average distance L before losing their identity and have an rms average velocity perpendicular to the wall of v^1, it can be readily shown that

$$E = \tfrac{1}{2} v^1 L = E_H = \varepsilon$$

If L is regarded as being identical for the three processes, one would thus expect the turbulent Prandtl number $Pr^1 = \varepsilon/E_H$ and the turbulent Schmidt number $Sc^1 = \varepsilon/E$ to be equal to unity. Actually, however, the transfer of momentum is not perfectly analogous to that of matter and heat, since in the case of momentum the transfer process is by virtue of the same property v. Hence, in fact, Pr^1 and Sc^1 are not unity, but they are both about equal to 0.70 and are independent of what diffuses (135).

In the transition zone, the equations will be

$$N_A = -(E + D) \frac{dc}{dy}$$

$$S = -(\mu + \varepsilon\rho) \frac{dU}{dy} \qquad (19\text{-}33)$$

$$q = -(k + E_H C_p\rho) \frac{dT}{dy}$$

If we define C_c, T_c, and U_c as the average concentration of diffusing material, the average temperature, and the average velocity across the fluid stream at any one cross section of the flow, respectively, then we can define the over-all mass- and heat-transfer coefficients and the friction factor f by the equations

$$k_c = \frac{N_{AW}}{C_c - C_u}$$

$$k_h = \frac{q_w}{T_c - T_w} \qquad (19\text{-}34)$$

$$f = \frac{S_w}{\frac{1}{2}\rho U_c{}^2}$$

where N_{AW}, q_w are the mass- and heat-transfer at the wall. Now N_A, q_w, and S_w are not constant across the cross section because each element of cross-sectional area carries some flow which contributes to the amount of matter, heat, and momentum available for radial transfer. Thus, Eqs. (19-31) to (19-33) cannot be simply integrated. If, however, one can assume, following Sherwood (114), that the variations of N_A, q_w, and S_w across the cross section are proportional to each other so that S/N_A and S/q are constant with respect to y, then one obtains equations of the form

$$\frac{dU}{dy} = \frac{S_w}{N_{AW}} \frac{E + D}{\mu + \varepsilon\rho} \frac{dC}{dy} = \frac{S_w}{q_w} \frac{k + E_H C_p\rho}{\mu + \varepsilon\rho} \frac{dT}{dy}$$

which can be integrated to give

$$U_c = \frac{S_w}{N_{AW}} \frac{E + D}{\mu + \varepsilon\rho} (C_c - C_w) = \frac{S_w}{q_w} \frac{k + E_H C_p\rho}{\mu + \varepsilon\rho} (T_c - T_w)$$

or $\qquad \dfrac{1}{\frac{1}{2}fU_c} = \dfrac{1}{k_c} \dfrac{E + D}{\mu + \varepsilon\rho} = \dfrac{1}{k_h} \dfrac{k + E_L C_p\rho}{\mu + \varepsilon\rho} \qquad (19\text{-}35)$

from Eq. (19-34).

For this integration to be valid, either the ratios $(E + D)/(\mu + \varepsilon\rho)$ and $(k + E_H C_p \rho)/(\mu + \varepsilon\rho)$ must remain constant from the wall, where E, E_H, and ε are zero, to the center of the stream, where E, E_H, and ε are much greater than $Dk/C_p\rho$ and μ/ρ, respectively, or else the transfer must be entirely governed by the region where all the transport is by eddies. These assumptions are neither of them fulfilled in general, so that instead of the simple equation (19-35) it is necessary to write

$$\frac{1}{\frac{1}{2}fU_c} = \frac{1}{k_c} f_1\left(\frac{\mu}{\rho D}, \frac{\varepsilon}{E}\right) = \frac{1}{k_h} f_2\left(\frac{C_p \mu}{k}, \frac{\varepsilon}{E_h}\right)$$

One would expect the functions f_1 and f_2 to be identical. Von Kármán (173) showed that an equation of this type is true for heat transfer during flow in pipes of any fluid, and Pigford (123) showed that it is true for mass transfer. Chilton and Colburn (120) obtained an empirical expression

$$\frac{1}{2}fU_c = k_c\left(\frac{\mu}{\rho D}\right)^{\frac{2}{3}} = k_h\left(\frac{C_p \mu}{k}\right)^{\frac{2}{3}}$$

Sherwood showed that for flow inside pipes, across flat plates, and around single cylinders these equations hold. This means that for systems of this type measurements of f as a function of the Reynolds number with one fluid can be used to determine k_c and k_h with any other fluid at the same Reynolds number in a system having the same geometry. It is clear, however, that the analogy between heat flow and mass flow is much closer than that between either of these properties and momentum flow or fluid friction. This follows because:

1. The ratios between the turbulent Schmidt and Prandtl numbers and between the laminar numbers are both nearly equal to 1.

2. Both matter and heat are carried by velocity, while momentum is carried by itself.

So far we have considered systems where the transport is to or from a solid boundary surface, but the story is very similar when one considers the transport of heat, matter, or momentum between two fluids away from walls. The case can be illustrated by that of the cylindrical free jet entraining a stagnant surrounding atmosphere. Hinze and van der Hegge Zijnen (107) have measured the turbulent mixing of a free jet with the surrounding stagnant atmosphere. In this system, there is conservation of momentum, heat, and nozzle fluid matter along the direction of the jet, and also the jet is conical. That is to say, the profiles of any property across the jet are similar at all distances (except distances less than a few diameters from the jet mouth), and the spread is proportional to the distance from the jet orifice. Under these conditions, the angle of spread of any property, or the angle between the axis of the jet and the cone of points where the value is one-half the value on the axis, is directly related to the reduction of this property along the axis by mixing because of the conservation of this property with distance. Hinze found that the momentum was transferred as if the jet had a constant eddy viscosity and that the concentration of nozzle fluid and the temperature excess of nozzle fluid over that of the surrounding fluid were transferred as if there were constant eddy diffusivity and eddy conductivity, respectively. However, the radial distributions of concentration and temperature were identical, falling to the half value at an angle of $\tan^{-1}$ 0.096, while the radial distribution of velocity fell to the half value at $\tan^{-1}$ 0.083. Thus, the turbulent Prandtl and Schmidt numbers are both about 0.74, and once again the analogy between mass flow and heat flow is much closer than that between either of these flows and momentum transfer.

These analogies have been used by various workers to calculate the rate of mass transfer in fuel-bed reactions between the gas stream and the particle surface. Either one can assume that the reaction rates are determined by the physical transport of matter and then use the analogy to predict the reaction, or one can observe the over-all reaction rate in an actual fuel bed and use the analogy to split the resistance to this reaction up into the physical and chemical resistances. Nusselt (163) was probably the first to use this type of analogy. Hougen and Watson (15) and Thring (66) have used the analogy between transfer of heat and matter to predict the combustion of oxygen in coke beds, making use of accepted figures from correlation of observations in different packed beds. Price and Thring (58) have separated the observed reaction rates between CO_2 and carbon into physical and chemical resistances by means of this analogy. Silver (63) has used the more indirect analogy between pressure drop and matter transfer. The analogies are not very reliable unless one has heat-transfer or pressure-drop measurements on the identical fuel bed, and this is the next step which is now being carried out. It is also necessary to take account of the difference in temperature between the reaction condition and the heat-transfer or pressure-drop system; this can readily be done if the temperatures are known, since the effect of temperature on the kinematic viscosity and on the Prandtl and Schmidt numbers for gases is well known. Where there are large temperature gradients within the fuel bed, the analogy will clearly be weakened, but as there is no possible method of measuring the matter transfer resistance directly, the analogy must perforce be used.

ANALOGUE MODELS FOR RADIANT-HEAT TRANSFER

These models may be divided into two types, those in which radiation of a different wavelength such as visible light is used to determine the distribution of thermal radiation, and those in which electric-analogue circuits are used for evaluating the radiant-heat-transfer equation.

Light Analogues for Thermal Radiation

In this system, a geometrically similar model of the furnace in which the heat transfer by radiation is to be evaluated is constructed on a greatly reduced scale. Heat-absorbing, relatively cold walls in the furnace are represented by black-painted surfaces or by the elimination of surfaces so that light leaves the model through these surfaces, while insulated refractory surfaces, which take up in the furnace a radiant equilibrium temperature and either reflect diffusely or absorb and reradiate the heat falling on them, are represented by white-painted surfaces, which give a diffuse reflection of light. The hot fuel bed or flame or electric source of radiation is represented by electric light bulbs behind an appropriate diffusing ground-glass screen. Heat flow in chu/(ft²)(hr) is represented in the model by light intensity in foot-candles. The amount of radiation falling on a given part of the heat-absorbing surface is measured by means of a photoelectric cell placed at the appropriate point. This procedure has been used by England and Croft (34) for the evaluation of the configuration factors for radiant-heat transfer between the hot square base of a rectangular furnace chamber and another wall when one or more surfaces are reflecting, i.e., represent insulated refractories. They found it necessary to allow for the fact that the photoelectric cell does not give a reading proportional to the cosine of the angle between the light direction and the normal to the cell when light from a point source falls on it at various angles. This correction was made by calibrating the cell with a point source of light placed at various angles. The cell was also calibrated for the effect of ambient temperature on its reading, and the true relationship between the output current and the illumination, which was not quite linear, was obtained by varying the distance of the point source. This technique has the obvious advantage of working with a model which is accurately geometrically similar to the real one and can therefore be very valuable in calculating systems which are too complicated in shape for mathematical evaluation of the radiation equations.

Electric Analogues for Radiation

Another method for evaluating the radiation equations in complicated systems was introduced by Paschkis (167). This method is somewhat similar to the use of the electric analogue for computing steady-state conduction, as it involves replacing the radiant-heat-transfer paths between the appropriate elements into which the system is subdivided by electric resistances and the replacement of sources of heat or surfaces maintained at a higher temperature by means of d-c batteries. It has the advantage over the light method that it is possible to bring in the effects of conduction heat losses from the refractory surface and convection heat transfer in addition to the radiation heat transfer. Conduction heat losses appear as a finite resistance connecting the inside surface to earth. It has, however, the very considerable disadvantage that the resistance equivalent to a radiation heat-transfer coefficient is strongly dependent on the temperatures of the two surfaces, so that the system has to be solved by successive approximations in which the resistances are changed to correspond to the voltages observed, and so the equivalent temperatures. The basic equations for radiation between any two black surfaces at absolute temperatures T_1, T_2 is as follows:

$$q_{df} = \frac{1}{\pi} \sigma \left[\left(\frac{T_1}{100} \right)^4 - \left(\frac{T_2}{100} \right)^4 \right] \int_{f_1} \int_{f_2} \frac{\cos \varphi_1 \cos \varphi_2}{r^2} df_1 \, df_2$$

where df_1, df_2 = elements of two areas

φ_1, φ_2 = angles between normals to these elements and the time joining them

σ = Stefan's constant

The heat transfer between two small elements of area can then be regarded as analogous to Ohm's law in the form

$$q_{df} = (t_1 - t_2) \frac{1}{R_{th}}$$

provided the thermal resistance of the heat-transfer circuit between the two elements of area has the value given by the equation

$$R_{th} = \frac{r^2}{\sigma \cos \varphi_1 \cos \varphi_2} \frac{\pi}{df_1 \, df_2} \frac{t_1 - t_2}{(T_1/100)^4 - (T_2/100)^4}$$

where t_1, t_2 = temperatures of surfaces, °C. This method of using the analogy is then to divide the radiating surfaces, the heat-absorbing surfaces, and the refractory surfaces into a number of areas, each of which can be taken as having a single temperature and for which the heat transfer can be regarded as occurring from the central point from which the angle φ can be measured. Alternatively, the areas can be somewhat larger if the integration of the configuration factors for the radiation between them has already been carried out mathematically. The equivalent electric circuit is then set up with a variable resistance for each heat-transfer path and fixed resistances for conducting paths. After observing the resulting temperatures of the intermediate points, the resistances can be readjusted to give the values of the radiation resistances corresponding to the observed temperatures, and, after a few steps of this successive approximation, a satisfactory solution of the radiation equations can be found. There are no condensers in the circuit because the problem is purely one of steady-state heat transfer. Paschkis applied this method to radiant heat transfer in electrically heated furnaces, and he was able to obtain the effect of heat losses through the walls and of the thermal resistance of the elements themselves.

More recently, the method has been applied by Oppenheim (166) to flame-heated systems and systems in which the space between the surfaces is filled with a gray gas and does not transmit radiation unimpeded. In the case of a gray gas within the system, this is equivalent to the appearance of an extra resistance between the surfaces, while the gas itself becomes a floating node, the potential of which, i.e., the temperature, depends on the heat transfer to it from the surroundings and its reradiation to the surroundings. In the case of combustion, the gas is itself a source of heat and therefore becomes not a floating node but a potential node. Since the electric circuits in these cases are confined to resistances and batteries, the electric-analogue computer obviously provides a very simple and convenient method of solving rather a complicated set of simultaneous equations and of allowing for the nonlinear radiation equations by a process of successive approximation.

ANALOGUES FOR STOCHASTIC PROCESSES

Stochastic processes, i.e., processes in which an input variable occurs in units which are random in time, are of very considerable importance in chemical engineering. As an example, suppose a works has a stockyard which is filled by unit supplies such as lorry- or trainload deliveries, and the arrival of these deliveries are random in time, that is to say, the chance of one arriving at a given moment is quite independent of how recently another one has arrived. Such a stockyard will in general be emptied at a uniform rate by the steadily proceeding process in the works, as, for example, the consumption of iron ore in a blast furnace. The problem in this case is to decide the most economical size of stockyard, i.e., the size which will give a sufficiently long time interval between two occasions on which the stockyard is completely empty without costing too much. Herne (139) has developed an analogue computer for this problem, based on the fact that the disintegration of a small radioactive source observes the same laws of complete randomness. A small radioactive source triggers off a Geiger counter at random intervals, and each pulse of the Geiger counter gives a standard electric charge to a condenser. The condenser is being discharged at a steady rate by a constant-current device, and the resulting potential of the condenser is observed. This will correspond to the extent to which the store is full. The storage condenser is also connected to a thyratron valve in such a way that if its potential falls to zero, corresponding to an empty store in the ruling case, the thyratron is fired and a point is recorded on a paper chart recorder. This firing also resets the condenser to a predetermined potential, corresponding to hurried partial refilling of the stockyard. Thus, the intervals between successive periods when the store is completely empty can be determined by allowing the system to run.

SYMBOLS IN CHAPTER 19

C = specific heat per unit mass, cal/(gm)(°C) or chu/(lb)(°C)

C_e = electrical capacity

C_p = specific heat per unit volume at constant pressure, cal/(cm³)(°C) or chu/(ft³)(°C)

c = concentration of diffusing material, lb moles/ft³

D = diffusion coefficient, cm²/sec or ft²/hr

E, ε, E_H = eddy diffusivities for matter, momentum, and heat, respectively, cm²/sec or ft²/hr

$\mathbf{E}$ = electric-field vector, volts/cm

f = friction factor, dimensionless

G = rigidity modulus

H = total energy of fluid per unit mass, cm²/sec²

$\mathbf{H}$ = magnetic-field vector

$\mathbf{i}$ = electric-current vector, amp/cm²

K = inductive capacity of dielectric

k = thermal conductivity, cal-cm/(cm²)(sec)(°C) or chu-ft/(ft²)(hr)(°C)

k_0 = thermal conductivity at $T = 0$

k_c = over-all mass-transfer coefficient, ft/hr

k_h = over-all heat-transfer coefficient, chu/(ft²)(hr)(°C)

M = magnetic permeability

N_A = mass-transfer rate, g moles/(sec)(cm²) or lb moles/(hr)(ft²)

P_r^1 = turbulent Prandlt number, ε/E_H

p = fluid pressure, dynes/cm²

q = vector flow of heat, cal/(cm²)(sec) or chu/(ft²)(hr)

R = volume density of electric charge

R_e = electrical resistance

S_x, S_y = shearing stresses in solid material, dynes/cm² or poundals/ft²

Sc^1 = turbulent Schmidt number ε/E

T = temperature, °C

t = time scale ratio between thermal prototype and electrical analogue

U_m, V_m = x and y components of flow velocity halfway between bounding planes

u, v, w = fluid-velocity components, cm/sec

V = electrostatic potential, volts

V^1 = fluid-force potential, cm²/sec²

X, Y, Z = external-force components, cm/sec²

Γ = specific electric resistance, ohms/cm²

γ_{xz} = xz component of shearing strain per unit length

θ = angle of rotation per unit length, radians/cm

μ = absolute fluid viscosity

ν = kinematic viscosity, cm²/sec or ft²/hr

ρ = fluid density, gm/cm³

Σ = shearing-stress function

σ = Stefan's constant, chu/(ft²)(hr)(°C⁴)

τ = tension on film per unit length, dynes/cm

ϕ = velocity function, cm²/sec

φ = angle

ψ = stream or current function, cm²/sec

Ω = magnetic potential

ω = vector space function (vorticity) with components ξ, η, ζ sec⁻¹

TERMINOLOGY

The application of model theory to chemical-engineering operations has made necessary the introduction of some new technical terms and the importation of others which may be unfamiliar to many chemical engineers. These terms are explained where they first occur in the text, but for convenience they are also listed below in alphabetical order. In introducing new terms, the authors have tried to make use of existing words rather than inventing new ones. This course has its disadvantages; for example, chemists and mathematicians may object to the special meaning given to the word "homologous" even though there is no danger of confusion with its chemical or geometrical meaning. The alternative would have been to invent a set of new words, thus creating yet another special jargon and making the book itself more difficult to read.

DEFINITION OF TERMS RELATING TO MODEL THEORY

analogue model An experimental apparatus in which certain physical variables are observed and the results used to predict the behavior of variables of a different kind in the prototype system which obey the same differential equations.

boundary effect A departure from similarity due to differing boundary conditions in model and prototype. Also called *wall effect.*

chemical regime A condition in which the over-all rate-determining process is a chemical-reaction velocity.

chemical similarity The condition of geometrically, kinematically, and thermally similar systems when corresponding concentration differences bear a constant ratio to one another. A special case under the general definition is when concentrations of reagents at corresponding points are equal.

corresponding concentration differences Concentration differences between corresponding pairs of points.

corresponding constituents Those constituents of chemically reacting systems with respect to which similarity is to be established.

corresponding forces Forces which act upon corresponding particles at corresponding times.

corresponding particles Geometrically similar particles centered upon corresponding points, the scale ratio of the particles being equal to that of the systems as a whole.

corresponding points Points situated in geometrically similar systems and such that the x, y, and z coordinates of one are, respectively, equal to L times the

x, y, and z coordinates in the other, L being the linear scale ratio. Hence also *corresponding lines* and *corresponding areas*.

corresponding temperature differences Temperature differences between corresponding pairs of points.

corresponding times Times, measured from a given datum, at which corresponding particles in a moving system have traced out geometrically similar lines, or in which temperatures have changed by corresponding amounts in unsteady-state heat-transfer systems. Hence also *corresponding time intervals*. The ratio of corresponding time intervals is the *time scale ratio* t.

corresponding velocities The velocities of corresponding particles at corresponding times.

critical operations Operations forming part of a pilot-scale process and for which the pilot plant is required to furnish plant design data.

dimensional constant Any constant in a physical equation which changes in value when the units of measurement are changed.

dimensional homogeneity The condition of a physical equation in which all terms have the same dimensions.

distorted model A model having different scale ratios along different axes. Hence also *distorted-model element*. See Fig. 3-2.

dynamic regime A condition in which the over-all rate-determining process is the physical dynamics of the system; in fluid systems, the flow pattern.

dynamic similarity The condition of geometrically similar moving systems when the ratios of corresponding forces are equal.

element A full-sized replica of one or more complete cells or components of the prototype apparatus (see Fig. 3-2).

empirical similarity A condition of geometrically similar systems in which conditions in the model are adjusted empirically to produce the desired degree of similarity to the prototype. It is then assumed that any change in the geometry of the model will produce a similar result to that given by a corresponding change in the geometry of the prototype.

extrapolation As applied to pilot plants and model experiments, extrapolation is the extension of similarity relations to systems in which conditions are not strictly similar. It is based upon the empirical assumption of power relations between variables.

flow pattern The geometrical form of the streamlines in a fluid-flow system.

generalized dimensionless equation An equation in which dimensionless groups are related by an unspecified function.

geometrical similarity The condition of two bodies or spaces when to every point in the one there exists a corresponding point in the other (see Fig. 3-1).

homologous systems Systems in which (1) the shapes of corresponding solid members or of the solid envelopes enclosing fluid masses are geometrically similar, (2) chemical compositions and physical properties at corresponding points (in so far as they affect the process being studied) are identical.

kinematic similarity The condition of geometrically similar systems in which corresponding particles trace out geometrically similar paths in corresponding intervals of time (see Fig. 3-3).

mass-action control Control of a chemical-reaction rate by mass action as in a homologous system.

mean residence time The mean residence time in a continuous flow vessel is taken as the volume of fluid in the vessel divided by the volumetric rate of flow.

mechanical similarity A generic term covering static, kinematic, and dynamic similarity.

mixed regime A condition in which an over-all rate of change is significantly influenced by two processes having incompatible similarity criteria.

model A geometrically similar replica of a larger apparatus termed the *prototype*. Unless otherwise stated, the scale ratio is assumed to be the same along three axes. See *distorted model*.

model element A scale model of an element of the prototype apparatus.

model experiment As contrasted with a pilot-plant test, a model experiment means an experiment using a model of an existing large-scale prototype. See *pilot plant*.

model law The scale-up law applicable to a given system.

model theory The theory of similarity as applied to scaling up or down.

pilot plant Any small-scale plant used to provide advance information about a future large-scale plant. The distinctions drawn by some authors between semitechnical, pilot, and semicommercial plants are disregarded here.

potential equation An equation in which a rate of change is equated to a potential difference divided by a resistance. Compound potential equations may be of series or parallel form according as the several resistances are in series or parallel.

prototype The large-scale apparatus to which a model or element is geometrically related. In this book the term *prototype* is used regardless of whether the large-scale apparatus comes into physical existence before or after the model. In other words, every model is considered to be related to a range of prototypes, either actual or hypothetical, and vice versa.

reduced velocity The actual fluid velocity divided by the lower critical velocity in the same system at the same temperature and pressure.

regime The nature of the over-all rate-determining process in a system.

residence time ratio (RTR) The ratio of mean to minimum residence time, or of maximum to mean fluid velocity, in a vessel through which there is continuous flow.

Reynolds index The exponent on the Reynolds number in the dimensionless rate equation for heat transfer, mass transfer, or momentum transfer (fluid friction) under conditions of forced convection. The Reynolds indices for heat, mass, and momentum transfer in a given system are not necessarily equal.

scale effect A departure from similarity on change of scale due to the existence of a mixed regime.

scale equations Simplified dimensionless equations which specify the conditions for similarity and the ratios of corresponding variables in terms of scale ratio for a given type of homologous system.

scale ratio The ratio of corresponding linear dimensions in prototype and model. In this book, the convention is that the prototype dimension is always the numerator of the ratio, whether the reference is to scaling up or down. Hence, scale ratios are always greater than unity. The symbol for scale ratio is L.

section ratio The ratio of corresponding cross-sectional areas in prototype and element, the convention being the same as for scale ratio. Hence, section ratios also are greater than unity. The symbol for section ratio is B^2.

similarity The condition of two or more systems in which there is a constant ratio between corresponding quantities. Similarity may be *geometrical, mechanical, thermal,* or *chemical* according to the properties in respect of which the systems are similar. Each of these degrees of similarity includes

all the previous ones.   Mechanical similarity is further subdivided into *static,*
kinematic, and *dynamic.*   Other kinds of similarity, such as *electrical,* are outside the scope of this book except in so far as they are employed in analogue
models.

slice model   A model element of a continuous-flow apparatus derived from a complete model by intersecting it with two planes parallel to the direction of flow.

static similarity (or **static-force similarity**)   The condition of geometrically similar
bodies or structures when under constant stress their relative deformations
are such that they remain geometrically similar.   The ratios of corresponding displacement are then equal to the scale ratio.

sticky-dust technique   A qualitative method of observing air flow over model surfaces by coating the surfaces with a sticky material and injecting a light powder into the air stream.   The distribution of powder on the surfaces varies
with the air-velocity distribution.

surface control   Control of a chemical-reaction rate by the rate of heat or mass
transfer across a surface or interface, as in many heterogeneous reactions.

symbols   The symbols used in each chapter are listed at the end of the chapter,
but there are two general conventions regarding corresponding quantities and
their ratios which should be mentioned here.   First, in symbols for corresponding quantities, the primed symbol always refers to the prototype.   Second, ratios of corresponding quantities are denoted by boldface type, it being
understood that the quantity relating to the prototype is the numerator of
the ratio.   Thus, $\mathbf{v} = v'/v$, the ratio of the velocity in the prototype to the
corresponding velocity in the model.   In Chap. 19 boldface type is also used
to denote vectors.

10° temperature coefficient   The rate of a process at a given temperature divided
by its rate at a temperature 10°C lower.   Strictly, the two temperatures
should be 25°/15°C.   The temperature coefficients of high-temperature processes may be corrected to 25°/15°C by means of the Arrhenius equation (see
Chap. 6).

thermal regime   A condition in which the over-all rate-determining process is a
rate of heat transfer.

thermal similarity   The condition of geometrically similar systems in which the
ratio of corresponding temperature differences is constant and which, if moving, are kinematically similar.   In thermally similar systems, the isotherm
patterns at corresponding times are geometrically similar.

time scale ratio   The ratio of corresponding times, that relating to the prototype
being the numerator of the ratio.   The symbol for time scale ratio is **t.**

wall effect   See *boundary effect.*

water model   A model in which the flow of air or gases in the prototype is simulated by water.

DIMENSIONS OF PHYSICAL AND CHEMICAL QUANTITIES*

Constant or variable	6-dimension system						4-dimension system			
	F	m	L	t	T	Q	m	L	t	T
Dimensional constants:										
Avogadro's number.................		-1					-1			
Chemical-equilibrium constant (ratio of two velocity constants).............										
Chemical-velocity constants:										
Homogeneous reactions:										
0 order........................		1	-3	-1			1	-3	-1	
1st order......................				-1					-1	
2d order.......................		-1	3	-1			-1	3	-1	
3d order.......................		-2	6	-1			-2	6	-1	
Heterogeneous reactions:										
0 order........................		1	-2	-1			1	-2	-1	
1st order......................			1	-1				1	-1	
2d order.......................		-1	4	-1			-1	4	-1	
3d order.......................		-2	7	-1			-2	7	-1	
Gas constant, energy units...........	1	-1	1		-1			2	-2	-1
Heat units........................		-1			-1	1		2	-2	-1
Henry's-law constant................	-1	1	-1					-3	2	
Mechanical equivalent of heat........	1		1			-1				
Newton's-law conversion factor.......	-1	1	1	-2						
Stefan-Boltzmann constant............			-2	-1	-4	1	1		-3	-4
Physical and chemical variables:										
Acceleration......................			1	-2				1	-2	
Action............................	1		1	1			1	2	-1	
Activity..........................										
Activity coefficient................										
Angle.............................										
Angular velocity...................				-1					-1	
Area..............................			2					2		
Bulk modulus......................	1		-2				1	-1	-2	

* Where no exponents are shown opposite a quantity, it has *zero* dimensions.

Constant or variable	6-dimension system						4-dimension system			
	F	m	L	t	T	Q	m	L	t	T
Compressibility	-1		2				-1	1	2	
Compressibility factor										
Concentration (mass/volume)		1	-3				1	-3		
Concentration gradient		1	-4				1	-4		
Density		1	-3				1	-3		
Diameter			1					1		
Diffusivity (diffusion coefficient)			2	-1				2	-1	
Efficiency										
Elastic modulus	1		-2				1	-1	-2	
Energy	1		1				1	2	-2	
Energy density	1		-2				1	-1	-2	
Enthalpy (unit mass)		-1				1		2	-2	
Entropy					-1	1	1	2	-2	-1
Equivalent weight		1					1			
Expansion coefficient, linear					-1					-1
Cubical					-1					-1
Fluidity (mass basis)		-1	1	1			-1	1	1	
Force	1						1	1	-2	
Free energy (unit mass)		-1				1		2	-2	
Frequency				-1					-1	
Friction coefficient, solid										
Friction factor, fluid										
Fugacity	1		-2				1	-1	-2	
Heat						1	1	2	-2	
Heat capacity (unit mass)		-1			-1	1		2	-2	-1
Heat capacity (unit volume)			-3		-1	1	1	-1	-2	-1
Heat content		-1				1		2	-2	
Heat of reaction		-1				1		2	-2	
Heat-transfer coefficient			-2	-1	-1	1	1		-3	-1
Internal energy (unit mass)		-1				1		2	-2	
Kinematic viscosity (mass basis)			2	-1				2	-1	
Latent heat		-1				1		2	-2	
Length			1					1		
Mass		1					1			
Mass-transfer coefficient, gas film	-1	1	1	-1				-1	1	
Liquid film			2	-1				2	-1	
Mass velocity		1	-2	-1			1	-2	-1	
Mole		1					1			
Mole fraction										
Molecular volume			-3					-3		
Molecular weight		1					1			
Moment	1		1				1	2	-2	
Moment, bending	1		1				1	2	-2	

Constant or variable	6-dimension system						4-dimension system			
	F	m	L	t	T	Q	m	L	t	T
Moment of inertia (area)			4					4		
Moment of inertia (mass)	1		1	2			1	2		
Moment of momentum	1		1	1			1	2	−1	
Momentum	1			1			1	1	−1	
Momentum flux	1						1	1	−2	
Power	1		1	−1			1	2	−3	
Pressure	1		−2				1	−1	−2	
Rate of flow (mass)		1		−1			1		−1	
Rate of flow (volume)			3	−1				3	−1	
Refractive index										
Solubility product		2	−6				2	−6		
Specific gravity										
Specific heat (unit mass)		−1			−1	1		2	−2	−1
Specific heat (unit volume)			−3		−1	1	1	−1	−2	−1
Specific surface			−1					−1		
Strain										
Stress	1		−2				1	−1	−2	
Surface tension	1		−1				1		−2	
Temperature					1					1
Temperature gradient			−1		1			−1		1
Tensile strength	1		−2				1	−1	−2	
Thermal conductance				−1	−1	1	1	2	−3	−1
Thermal conductivity			−1	−1	−1	1	1	1	−3	−1
Thermal diffusivity			2	−1				2	−1	
Thermal-expansion coefficient, linear						−1				−1
Cubical						−1				−1
Thermal resistance				1	1	−1	−1	−2	3	1
Time				1					1	
Torque	1		1				1	2	−2	
Velocity			1	−1				1	−1	
Angular				−1					−1	
Viscosity (force basis)	1		−2	1			1	−1	−1	
Viscosity (mass basis)		1	−1	−1			1	−1	−1	
Volume			3					3		
Weight	1						1	1	−2	
Work	1		1				1	2	−2	
Young's modulus	1		−2				1	−1	−2	

APPENDIX 3

DIMENSIONLESS GROUPS

Arrhenius . . . . $\dfrac{E}{RT}$

Biot . . . . $\dfrac{hL}{k}$

Condensation . . . . $\dfrac{h}{k}\left(\dfrac{\nu^2}{a}\right)^{1/3}; \dfrac{h}{k}\left(\dfrac{\nu^2}{g}\right)^{1/3}$

Euler . . . . $\dfrac{g_c p}{\rho v^2}$

Fourier . . . . $\dfrac{kt}{\rho c L^2}$

Froude . . . . $\dfrac{v^2}{gL}$

Graetz . . . . $\dfrac{wc}{kL}$

Grashof . . . . $\dfrac{L^3 \rho^2 \beta g \ \Delta T}{\mu^2}$

Mach . . . . $\dfrac{v}{v_a}$

Nusselt . . . . $\dfrac{hd}{k}$

Peclet . . . . $\dfrac{v \rho c L}{k}$

Prandtl . . . . $\dfrac{c\mu}{k}$

Power (for mixing) . . . . $\dfrac{P g_c}{\rho N^3 d^5}$

Pressure coefficient . . . . $\dfrac{F}{\rho v^2 d^2} = \dfrac{\Delta p}{\rho v^2}$

Reynolds . . . . $\dfrac{\rho v L}{\mu} = \dfrac{vL}{\nu}$

Schmidt . . . . $\dfrac{\mu}{\rho D}$

Sherwood . . . . $\dfrac{k_c L}{D}$

Stanton . . . . $\dfrac{h}{cv\rho}$

Thring (radiation) . . . . $\dfrac{\rho c v}{se T^3}$

Vapor condensation. . . . . $\dfrac{L^3\rho^2 g\lambda}{k\mu\,\Delta T}$

Weber. $\dfrac{\rho v^2 L}{\sigma}$

SYMBOLS USED IN DIMENSIONLESS GROUPS

a = acceleration
c = specific heat
D = diffusivity
d = diameter
E = activation energy
e = emissivity
F = force
g = acceleration of gravity
g_c = Newton's-law conversion factor
h = heat-transfer coefficient, individual
k = thermal conductivity
k_c = mass-transfer coefficient, individual
L = length
P = power consumption
p = pressure
R = gas constant
s = Stefan-Boltzmann constant
T = temperature
t = time
v = linear velocity
v_a = velocity of sound in fluid
w = mass-flow rate
β = coefficient of cubical expansion
Δ = difference (as in Δp, ΔT)
λ = latent heat of vaporization
μ = viscosity
ν = kinematic viscosity
ρ = density
σ = surface or interfacial tension

APPENDIX 4

NOTE ON AUTOMATIC CONTROL

The object of this note is not to discuss automatic control as such, or even to make suggestions about the instrumentation of pilot plants, but to draw attention to the possibility of predicting the control characteristics of full-sized process plants by means of experiments with models. So far as is known, this technique has not yet been applied in practice.

The transient dynamics of a process plant are characterized by the time lag between the occurrence of a change in the controlled variable and the application of a correction by the controller. This time lag is made up of four components: distance-velocity lag and transfer lag in the plant, measuring lag and transmission lag in the instrument system. Briefly distance-velocity lag is the time interval between a sudden disturbance on the input side and the first appearance of an effect on the output side. Transfer lag is a capacity effect which transforms the sudden input disturbance into a more or less gradual change in the output stream. Measuring lag is the delay before the change in output condition is picked up by the detecting element (due, for example, to the thermal capacity of the thermometer pocket). Transmission lag is the time taken for the detecting impulse to reach the controller and for the motor impulse to reach the control valve. In the case of pneumatic systems, it may be appreciable.

Measuring and transmission lags are properties of the instrument systems which are usually known or calculable. Distance-velocity lag as such is calculable also; in general, it represents the time taken by the process fluid to travel from the point at which the disturbed condition is corrected to the point at which it is measured. Distance-velocity lag, however, is always associated with transfer lag, and the latter is not at present calculable except for the simplest types of flow system. Without a knowledge of the transfer lag, it is impossible to select suitable controller characteristics.

Current practice is to install industrial control instruments that are adjustable over a wide range, determine the dynamics of the plant to be controlled by means of a full-scale test, and finally adjust the proportional band width, reset rate, etc., to suitable values. There are two methods of making such full-scale tests, known, respectively, as the *step-response method* and the *frequency-response method*. In the step-response method, the controller is temporarily put out of action, a sudden change is artifically imposed upon the input variable, and the change in the output variable is measured and plotted against time. From the resulting curve, the optimum settings can be determined. The step-response method is simple in theory but difficult to apply in practice because the response curve is liable to be distorted by random variations.

In the frequency-response method, a comparatively small sinusoidal pulsation is impressed upon the input variable, and the resulting pulsation in the output

286

variable is recorded. For example, in a steam-heated tubular water heater, the pneumatic valve controlling the steam input might be caused to pulsate by a sinusoidal-wave generator and the consequent pulsations in the hot-water temperature observed. Two parameters are calculated, the phase lag between the input and output pulsations and the "attenuation," i.e., the ratio of input to output amplitude. From these two quantities, determined over a range of frequencies, the control characteristics of the plant can be deduced.

A test by the step-response method is recorded as a pair of magnitude-time curves, one showing the sudden "step" in the input variable, and the other the more gradual change in the output variable. The record of a frequency-response test is a graph of the sinusoidal input and output curves, showing their respective amplitudes and the time lag or phase angle between them.

It is not necessary to describe here how these response curves are used to evaluate controller characteristics. The experimental procedures and methods of calculation are given in standard textbooks. The present question is whether, instead of waiting until the full-scale plant is built and in operation to determine instrument characteristics, step-response or frequency-response tests could be carried out on a model and the results used to predict the control dynamics of the prototype plant.

Consider the steam-heated tubular water heater mentioned earlier, and let there be a scale model or model element operated under conditions of dynamic similarity. In the prototype, a sudden rise in steam pressure causes an increase in water temperature which reaches the outlet temperature-measuring element in a time t, depending on the water velocity. In the model, if L be the scale ratio (prototype/model), the distance from heating surface to measuring element is $1/L$ times that in the prototype, and the corresponding velocity for similarity is L times as great. Hence, the time for a sudden temperature increase in the model to reach the measuring elements is

$$\frac{t}{L^2}$$

A step-response graph for the model could therefore be converted into a graph for the prototype by multiplying the time scale by L^2.

Similar considerations apply to a frequency-response graph. The corresponding frequency in the prototype would be $1/L^2$ times that in the model. At these frequencies, model and prototype systems might be expected to have the same phase lag and attenuation.

Where the two experimental systems are not homologous, as when a water model is used to simulate a large air preheater, the same principle should apply; but here the time scale of the model response curve should be multiplied by

$$\frac{L^2}{v}$$

where v is the prototype/model ratio of kinematic viscosities.

It is often impracticable to operate a model under conditions of dynamic similarity, especially where the systems are homologous. As was shown in Chap. 8, there are empirical methods of extrapolation which can be applied to steady-state variables in such cases. Probably empirical methods of extrapolating step-response and frequency-response curves could also be developed from suitable data.

Where pilot-plant units are designed and operated in accordance with model theory, as has been advocated in this book, it should theoretically be possible to obtain information from them about control dynamics and perhaps even fix controller characteristics before the full-scale plant is built.   Such methods could not be applied with any confidence, however, until they have been verified experimentally.   Here is virgin soil for research: the scaling up of process-control systems.

CLASSIFIED REFERENCES

The published works referred to in the text are here classified by subjects. The subject headings are intentionally different from the chapter headings of the text, since the same paper is often mentioned in more than one chapter. In this classified list, no reference appears more than once; if a paper deals with several subjects, it is listed under that of greatest interest to the present work. It is hoped that with this arrangement the list of references will also be of use as a bibliography of literature having some bearing on model theory as applied to the various branches of chemical engineering.

CHEMICAL REACTIONS

1. Bosworth, R. C. L.: *J. Roy. Soc. New South Wales*, **81**:15 (1947).
2. Bosworth, R. C. L.: *Trans. Faraday Soc.*, **53**:399 (1947).
3. Bosworth, R. C. L.: *Phil. Mag.*, **39**:847 (1948).
4. Bosworth, R. C. L.: *Phil. Mag.*, **40**:314 (1949).
5. Brinkley, S.: *Ind. Eng. Chem.*, **40**:303 (1948).
6. Brothman, A., A. P. Weber, and Z. Barish: *Chem. Met. Eng.*, **50**(7):111, **50**(8):107, **50**(9):113 (1943).
7. Damköhler, G.: *Z. Elektrochem.*, **42**:846 (1936).
8. Damköhler, G.: *Chem. Fabrik*, **12**:469 (1939).
9. Danckwerts, P. V., J. W. Jenkins, and G. Place: *Chem. Eng. Sci.*, **3**:26 (1954).
10. Denbigh, K. G.: *Trans. Faraday Soc.*, **40**:352 (1944).
11. Denbigh, K. G.: *J. Appl. Chem. (London)*, **1**:227 (1951).
12. Dodd, R. H., and K. M. Watson: *Trans. AICE*, **42**:263 (1946).
13. Dodd, R. H.: *Intern. Congr. Pure and Appl. Chem.*, 1947.
14. Eldridge, J. W., and E. L. Piret: *Chem. Eng. Progr.*, **46**:290 (1950).
15. Hougen, O. A., and K. M. Watson: *"Chemical Process Principles,"* vol. 3, p. 1068, John Wiley & Sons, Inc., New York, 1947.
16. Hulburt, H. M.: *Ind. Eng. Chem.*, **36**:1012 (1944).
17. Hurt, D. M.: *Ind. Eng. Chem.*, **35**:522 (1943).
18. Johnstone, R. Edgeworth: *Trans. Inst. Chem. Engrs. (London)*, **17**:129 (1939).
19. Jones, R. W.: *Chem. Eng. Progr.*, **47**:46 (1951).
20. Kirillov, N. I.: *J. Appl. Chem. (U.S.S.R.)*, **13**:978 (1940).
21. Laupichler, F. G.: *Ind. Eng. Chem.*, **30**:578 (1938).
22. McAllister, S. H.: *Oil Gas J.*, Nov. 12, 1937, p. 139.
23. MacMullin, R. B.: *Chem. Eng. Progr.*, **44**:183 (1948).
24. MacMullin, R. B., and M. Weber: *Trans. AIChE*, **31**:409 (1935).
25. Thiele, E. W.: *Ind. Eng. Chem.*, **31**:916 (1939).

COMBUSTION

26. Baron, T.: *Chem. Eng. Progr.*, **50**:73 (1954).
27. Bennett, J. G., and R. L. Brown: *J. Inst. Fuel*, **13**:232 (1940).
28. Bishop, T.: *Iron & Coal Trades Rev.*, **155**:835 (Oct. 31, 1947).
29. British Iron & Steel Research Association (Physics Department): private communication.
30. Chesters, J. H., R. S. Howes, I. M. D. Halliday and A. R. Philip: *J. Iron Steel Inst. (London)*, **162**:385 (1949).
31. Collins, R. D., and M. P. Newby: *Nature*, **162**:224 (Aug. 7, 1948).
32. Collins, R. D., and J. D. Tyler: *J. Iron Steel Inst. (London)*, **162**:457 (1949).
33. Dunningham, A. C., and E. S. Grummell: *J. Inst. Fuel*, **11**:117, 129 (1937).
34. England, F., and H. O. Croft: *Trans. ASME*, **64**:691 (1942).
35. Flame Radiation Joint Research Committee: *J. Inst. Fuel Rept.*, **24**:51 (1951)
36. Flame Radiation Joint Research Committee: *J. Inst. Fuel*, **25**:517 (1952).
37. Flame Radiation Joint Research Committee: *J. Inst. Fuel*, **26**:189 (1953).
38. Gaydon, A. G., and H. G. Wolfhard: "Flames," Chapman & Hall, Ltd., London, 1953.
39. Gooding, E. J., and M. W. Thring: *Trans. Soc. Glass Technol.*, **25**:21 (1941).
40. Groume-Grjimailo, W. E.: "The Flow of Gases in Furnaces," John Wiley & Sons, Inc., New York, 1923.
41. Horn, G., and M. W. Thring: private communication.
42. Hughes, M. L.: *J. Iron Steel Inst. (London)*, **156**:55 (1947).
43. Hughes, M. L.: *J. Iron Steel Inst. (London)*, **156**:371 (1947).
44. Leckie, A. H., et al.: *J. Iron Steel Inst. (London)*, **155**:392 (1947).
45. Leckie, A. H., et al.: *J. Iron Steel Inst. (London)*, **160**:49 (1948).
46. Leigh, E. T.: *Brit. Iron Steel Research Assoc. Document P/C* 191, 1953.
47. Leys, J. A.: *Research*, **5**:439 (September, 1952).
48. Leys, J. A., and E. T. Leigh: *J. Iron Steel Inst. (London)*, **165**:301 (1950).
49. Leys, J. A., and E. T. Leigh: *J. Iron Steel Inst. (London)*, **170**:336 (1952).
50. Mache, H.: *Forsch. geirete Ingenieurw*, **14**:77 (1943).
51. Marskell, W. G., and J. M. Miller: *Fuel*, **25**:4 (1946).
52. Marskell, W. G., J. M. Miller and W. I. Joyce: *Fuel*, **31**:91 (1952).
53. Newby, M. P.: *J. Iron Steel Inst. (London)*, **162**:452 (1949).
54. Newby, M. P., R. D. Collins and J. A. Leys: *Brit. Iron Steel Research Assoc. Document P/C* 68.
55. Newby, M. P., and S. E. McEwan: *Brit. Iron Steel Research Assoc. Document P/C* 99.
56. Nicholls, P.: *U. S. Bur. Mines Bull.* 378, 1934.
57. Pechès, I.: *Verres et réfractaires*, June, 1947, p. 3.
58. Price, P. H., and M. W. Thring: "*International Conference on Complete Gasification of Coal*," p. 142, Inichar Liége, 1954.
59. Rosin, P. O.: *J. Inst. Fuel*, **9**:287 (1936).
60. Rosin, P. O.: The Flow of Air in a Gas Producer, Joint Report of the British Iron & Steel Research Association and the British Coal Utilisation Research Association, June, 1939.
61. Saunders, H. L., and R. Wild: *J. Iron Steel Inst. (London)*, **152**:259 (1945).
62. Schild, A.: *Glastech. Ber.*, **11**:305 (1933).
63. Silver, R. S.: *Fuel*, **32**:121 (1953).
64. Thring, M. W.: *Research*, **1**:492 (1948).

65. Thring, M. W.: *Research*, **2**:36 (1949).
66. Thring, M. W.: *Fuel*, **31**:355 (1952).
67. Thring, M. W.: "The Science of Flames and Furnaces," Chapman & Hall, Ltd., London, 1952.
68. Thring, M. W., and M. P. Newby: *Proc. 4th Intern. Symposium Combustion*, Cambridge, Mass., September, 1952, p. 789.
69. Traustel, S.: *Iron & Coal Trades Rev.*, **158**:1167, 1225 (1949).
70. Voice, E. W., C. Lang, and P. K. Gledhill: *J. Iron Steel Inst. (London)*, **167**: 393 (1951).
71. Watson, E. A., and J. S. Clarke: *J. Inst. Fuel*, **21**:2 (October, 1947).

CORROSION

72. Evans, U. R.: numerous papers summarized in "Metallic Corrosion, Passivity and Protection," 2d ed., Edward Arnold & Co., London, 1948.
73. Inglesent, H., and J. A. Storrow: *J. Soc. Ind. Chem.*, **64**:233 (1945).
74. Müller, W. J., and K. Konopicky: *Sitzuber. Akad. Wiss. Wien. Math.-naturw. Kl.*, Abt. IIb, 138 suppl., p. 707, 1929.
75. Speller, F. N., and V. V. Kendall: *Ind. Eng. Chem.*, **15**:134 (1923).

CRUSHING AND GRINDING

76. Bachmann, D.: unpublished work, 1937, quoted by W. Matz (252).
77. Coghill, W. H., and F. D. DeVaney: *U. S. Bur. Mines Tech. Paper* 581, 1937.
78. Davis, E. W.: *Trans. AIME*, **61**:250 (1919).
79. Fischer, H.: *Z. Ver. deut. Ing.*, **48**:437 (1904).
80. Gow, A. M., A. B. Campbell, and W. H. Coghill: *Trans. AIME*, **87**:51 (1930).
81. Gow, A. M., M. Guggenheim, A. B. Campbell, and W. H. Coghill: *Trans. AIME*, **112**:24 (1935).
82. Michaelson, S. D.: *AIME Tech. Publ.* 1844, 1945.

DISTILLATION AND EVAPORATION

83. Duncan, D. W., J. H. Koffolt, and J. R. Withrow: *Trans. AIChE*, **38**:259 (1942).
84. Furnas, C. C., and M. L. Taylor: *Trans. AIChE*, **36**:135 (1940).
85. Gamson, B. W., G. Thodos, and O. A. Hougen: *Trans. AIChE*, **39**:1 (1943).
86. Hands, C. H. G., and F. R. Whitt: *J. Appl. Chem.*, Mar. 1, 1951, p. 135.
87. Jacobs, J.: "Distillier-Rektifizier-Anlagen," R. Oldenbourgh-Verlag, Munich, 1950.
88. Kirschbaum, E.: "Distillation and Rectification," Chemical Publishing Company, Inc., New York, 1948.
89. Peters, W. A.: *Ind. Eng. Chem.*, **14**:476 (1922).
90. Robinson, C. S., and E. R. Gilliland: "Elements of Fractional Distillation," 4th ed., McGraw-Hill Book Company, Inc., New York, 1950.
91. Wilke, C. R., and O. A. Hougen: *Trans. AIChE*, **41**:445 (1945).
See also *Packed Towers (Flow Pattern)*.

FILTRATION

92. Almy, C., and W. K. Lewis: *Ind. Eng. Chem.*, **4**:528 (1912).
93. Carman, P. C.: *Trans. Inst. Chem. Engrs.* (*London*), **16**:168 (1938).
94. McMillen, E. L., and H. A. Webber: *Trans. AIChE*, **34**:213 (1938).
95. Ruth, B. F.: *Ind. Eng. Chem.*, **38**:564 (1946).
96. Ruth, B. F., and L. L. Kempe: *Trans. AIChE*, **33**:34 (1937).
97. Ruth, B. F., G. H. Montillon, and R. E. Montonna: *Ind. Eng. Chem.*, **25**:76, 153 (1933).
98. Sperry, D. R.: *Ind. Eng. Chem.*, **36**:323 (1944).
99. Underwood, A. J. V.: *Trans. Inst. Chem. Engrs.* (*London*), **4**:19 (1926).

FLOW OF FLUIDS AND SOLIDS

100. Beale, E. S. L., and P. Docksey: "Technical Data on Fuel," 5th ed., p. 67, World Power Conference (British National Committee), 1950
101. British Standard 1042: Flow Measurement, p. 60, 1943.
102. Chesters, J. H., and M. W. Thring: *Iron Steel Inst. Special Rept.* **37**, p. 38, 1946.
103. Durand, R., and E. Condolios: IIme journée de l'hydraulique, *Compt. rend.*, June, 1952, p. 29, Société hydrotechnique de France, Paris.
104. Friedman, S. J., F. A. Gluckert, and W. R. Marshall: *Chem. Eng. Progr.*, **48**: 181 (1952).
105. Genereaux, R. P.: *Ind. Eng. Chem.*, **29**:385 (1937).
106. Green, A. P.: *Phil. Mag.*, **42**:365 (1951).
107. Hinze, H. O. E., and B. G. van der Hegge Zijnen: *Appl. Sci. Research*, **A1**: 435 (1947–1949).
108. Leys, J. A.: *Research*, **5**:439 (1952).
109. Needham, H. C.: *Power Jets* (*Research & Development*), Ltd. Rept. R. 1209.
110. Nelson, L. C., and E. F. Obert: *Chem. Engineering*, **61**:213 (July, 1954).
111. Obert, E. F.: *Ind. Eng. Chem.*, **40**:2185 (1948).
112. Radcliffe, A.: *Proc. Inst. Mech. Engrs.* (*London*), **169**:93 (1955).
113. Relf, E. F.: *Phil. Mag.*, **48**:535 (1924).
114. Sherwood, T. K.: *Ind. Eng. Chem.*, **42**:2077 (1950).
115. Spells, K. E.: *Trans. Inst. Chem. Engrs.* (*London*), **33**:79 (1955).
116. Stanton, T. E., and J. R. Pannell: *Trans. Roy. Soc.* (*London*), **A214**:199 (1914).
117. Walton, W. H., and W. C. Prewett: *Proc. Roy. Soc.* (*London*), **62**(354B):341 (1949).
118. Worster, R. C.: *Proc. Colloq. Hydraulic Transport Coal* (National Coal Board), London, Nov. 5, 6, 1952.
See also *Packed Towers* (*Flow Pattern*).

GAS ABSORPTION

119. Ayerst, R. P., and L. S. Herbert: *Trans. Inst. Chem. Engrs.* (*London*), **32** (suppl.):568 (1954).
120. Chilton, T. H., and A. P. Colburn: *Ind. Eng. Chem.*, **26**:1183 (1934).

121. Emmert, R. E., and R. L. Pigford: *Chem. Eng. Progr.*, **50**:87 (1954).
122. Morris, G. A., and J. Jackson: "Absorption Towers," Butterworth & Co. (Publishers) Ltd., 1953.
123. Pigford, R. L.: private communication, 1948.
124. Pratt, H. R. C.: *Trans. Inst. Chem. Engrs. (London)*, **29**:195 (1951).
125. Scheibel, E. C., and D. F. Othmer: *Trans. AIChE*, **40**:611 (1944).
126. Sherwood, T. K., and F. A. L. Holloway: *Trans. AIChE*, **36**:21, 39 (1940).
127. Sherwood, T. K., and R. L. Pigford: "Absorption and Extraction," 2d ed., McGraw-Hill Book Co., Inc., New York, 1952.
128. Wiegand, J. H.: doctorate research, University of Michigan. Quoted in J. H. Perry (ed.), "Chemical Engineers' Handbook," 2d ed., p. 1156, McGraw-Hill Book Company, Inc., New York, 1941. Does not appear in 3d edition.
See also *Packed Towers (Flow Pattern)*.

GENERAL AND MISCELLANEOUS

129. Brownlee, K. A.: "Industrial Experimentation," H.M. Stationery Office London, 1st ed., 1946; 3d ed., 1948.
130. Carley, J. F., and J. M. McKelvey: *Ind. Eng. Chem.*, **45**:985 (1953).
131. Carley, J. F., and R. A. Strub: *Ind. Eng. Chem.*, **45**:970 (1953).
132. Davies, O. L. (ed.): "Statistical Methods in Research and Production," Oliver & Boyd, Ltd., Edinburgh and London, 1949.
133. Davies, O. L. (ed.): "The Design and Analysis of Industrial Experiments," Oliver & Boyd, Ltd., Edinburgh and London, 1954.
134. Davis, George E.: "A Handbook of Chemical Engineering," Davis Bros., Manchester, England, 1901.
135. Forstall, W., and A. H. Shapiro: *Mass. Inst. Technol. Meteorol. Rept.* 39, 1949.
136. Glasstone, S., K. L. Laidler and H. Eyring: "The Theory of Rate Processes," McGraw-Hill Book Company, Inc., New York, 1941.
137. Gore, W. L.: *Ind. Eng. Chem.*, **43**:2327 (1951).
138. Hele-Shaw, H. S., and A. Hay: *Trans. Inst. Naval Arch.*, **40** (1898); *Trans. Roy. Soc. (London)*, **A195**:303 (1901).
139. Herne, H.: *Proc. Isotope Techniques Conf.*, (Oxford, **11**: 1951).
140. Jeans, J. H.: "Mathematical Theory of Electricity and Magnetism," Cambridge University Press, New York, 5th ed., p. 41, 1925.
141. Perry, J. H. (ed.): "Chemical Engineers' Handbook," 3d ed., McGraw-Hill Book Company, Inc., New York, 1950, especially sec. 8, General Theory of Diffusional Operations, by A. P. Colburn and R. L. Pigford.
142. Stokes, G. G.: "Encyclopedia Britannica," 11th ed., Vol. **14**, p. 38, 1910.
143. Walker, W. H., W. K. Lewis, W. H. McAdams, and E. R. Gilliland: "Principles of Chemical Engineering," 3d ed., McGraw-Hill Book Co., Inc., New York, 1937.
144. Yates, F.: *Nature*, **170**:138 (1952).

HEAT TRANSFER

145. Andrews, R. V.: *Chem. Eng. Progr.*, **51**:67F (1955).
146. Awberry, J. H. and F. H. Schofield: *Proc. 5th Intern. Congr. Refrig.*, **3**:591 (1929).

147. Beuken, C. L.: *Econom. Tech. Tidschr.*, **19**(3):43 (1949).
148. Chilton, T. H., T. B. Drew and R. H. Jebens: *Ind. Eng. Chem.*, **36**:510 (1944).
149. Coyle, M. B.: General Discussion on Heat Transfer, *Inst. Mech. Engrs. (London)*, 1951, p. 265.
150. Denbigh, K. G.: *J. Soc. Chem. Ind. (London)*, **65**:61 (1946).
151. Fishenden, M., and O. A. Saunders: "An Introduction to Heat Transfer," Oxford University Press, New York, 1950.
152. Insinger, T. H., and H. Bliss: *Trans. AIChE*, **46**:491 (1940).
153. Jakob, M.: *Z. ges. Kälte-Ind.*, **29**:83 (1922)
154. Jakob, M.: *Proc. 5th Intern. Congr. Appl. Mech.*, 1938, p. 561.
155. Jakob, M.: "Heat Transfer," p. 402, John Wiley & Sons, Inc., New York, 1949.
156. Johnstone, R. Edgeworth: *Trans. Inst. Chem. Engrs. (London)*, **33**:243 (1954).
157. Kayan, C. F.: *Trans. ASME*, **67**:713 (1945).
158. Kron, G.: *J. Appl. Phys.*, **6**:172 (1945).
159. Langmuir, I., E. Q. Adams, and F. S. Meikle: *Trans. Am. Electrochem. Soc.*, **24**:53 (1913).
160. Lorenz, L.: *Wied. Ann.*, **13**:562 (1881).
161. McAdams, W. H.: "Heat Transmission," 3d ed., McGraw-Hill Book Company, Inc., New York, 1955.
162. Moore, A. D.: *Ind. Eng. Chem.*, **28**:704 (1936).
163. Nusselt, W.: *Z. Ver. deut. Ing.*, **60**:102 (1916).
164. Nusselt, W.: *Z. Ver. deut. Ing.*, **60**:541, 569 (1916).
165. Oldshue, J. Y., and A. T. Gretton: *Chem. Eng. Progr.*, **50**:615 (1954).
166. Oppenheim, A. K.: *Shell Development Co. Tech. Rept.* 246-52 (1952).
167. Paschkis, V.: *Elektrotech. u. Maschinenbau*, **54**:617 (1936).
168. Paschkis, V.: *Trans. ASME*, **38**:117 (1947).
169. Paschkis, V., and H. D. Baker: *Trans. ASME*, **33**:105 (1942).
170. Pratt, N. H.: *Trans. Inst. Chem. Engrs. (London)*, **25**:163 (1947).
171. Saunders, O. A., and H. Ford: *J. Iron Steel Inst.*, **141**:291P (1940).
172. Seider, E. N., and G. E. Tate: *Ind. Eng. Chem.*, **28**:1429 (1936).
173. Von Kármán, T.: *Trans. ASME*, **61**:705 (1939).
174. Wilson, E. E.: *Trans. ASME*, **37**:47 (1915).
175. Wilson, W. I., and A. J. Miles: *J. Appl. Phys.*, **21**:532 (1950).

LIQUID EXTRACTION

175a. Baron, T.: private communication.
176. Colburn, A. P., and D. G. Welsh: *Trans. AIChE*, **38**:179 (1942).
177. Crawford, J. W., and C. R. Wilke: *Chem. Eng. Progr.*, **47**:423 (1951).
178. Dell, F. R., and H. R. C. Pratt: *Trans. Inst. Chem. Engrs. (London)*, **29**:89 (1951).
179. Gayler, R., and H. R. C. Pratt: *Trans. Inst. Chem. Engrs. (London)*, **31**:69, 78 (1953).
180. Hou, H. L., and N. W. Franke: *Chem. Eng. Progr.*, **45**:65 (1949).
181. Laddha, G. S., and J. M. Smith: *Chem. Eng. Progr.*, **46**:195 (1950).
182. Lewis, J. B., I. Jones, H. R. C. Pratt: *Trans. Inst. Chem. Engrs. (London)*, **29**:126 (1951).

182a. Pratt, H. R. C.: private communication.
183. Pratt, H. R. C., and S. T. Glover: *Trans. Inst. Chem. Engrs. (London)*, **24**:54 (1946).
184. Sherwood, T. K., J. E. Evans, and J. V. A. Longcor: *Trans. AIChE*, **35**:597 (1939).
185. Treybal, R. E.: "Liquid Extraction," McGraw-Hill Book Company, Inc., New York, 1951.
See also *Packed Towers (Flow Pattern)*.

MIXING

186. Bissell, E. S., F. D. Miller, and Everett: *Ind. Eng. Chem.*, **37**:426 (1945).
187. Brothman, A., and H. Kaplan: *Chem. Met. Eng.*, **46**:633, 639 (1939).
188. Buche, W.: *Z. Ver. deut. Ing.*, **81**:1065 (1937).
189. Danckwerts, P. V.: *Chem. Eng. Sci.*, **2**, 2 (1953).
190. Green, S. J.: *Trans. Inst. Chem. Engrs. (London)*, **31**:327 (1953).
191. Hixson, A. W., and S. J. Baum: *Ind. Eng. Chem.*, **33**:478 (1941).
192. Hixson, A. W., and S. J. Baum: *Ind. Eng. Chem.*, **34**:120 (1942).
193. Hixson, A. W., and J. H. Crowell: *Ind. Eng. Chem.*, **23**:923, 1002, 1160 (1931).
194. Hixson, A. W., and V. D. Luedeke: *Ind. Eng. Chem.*, **29**:927 (1937).
195. Hixson, A. W., and M. I. Smith: *Ind. Eng. Chem.*, **41**:973 (1949).
196. Hixson, A. W., and A. H. Tenney: *Trans. AIChE*, **31**:113 (1935).
197. Hixson, A. W., and G. A. Wilkins: *Ind. Eng. Chem.*, **25**:1196 (1933).
198. Johnstone, R. Edgeworth, *Trans. Inst. Chem. Engrs. (London)*, **31**:327 (1953). Discussion.
199. Marc, K.: *Z. physik. Chem.*, **61**: 385 (1908).
200. Marc, K.: *Z. physik. Chem.*, **67**:470 (1909).
201. Marc, K.: *Z. physik. Chem.*, **68**:104 (1909).
202. Marc, K.: *Z. physik. Chem.*, **73**:685 (1910).
203. Martin, J. J.: *Trans. AIChE*, **42**:777 (1946).
204. Miller, S. A., and C. A. Mann: *Trans. AIChE*, **40**:709 (1944).
205. Olney, R. B., and G. J. Carlson: *Chem. Eng. Progr.*, **43**:473 (1947).
206. Rushton, J. H.: *Ind. Eng. Chem.*, **37**:422 (1945).
207. Rushton, J. H.: *Chem. Eng. Progr.*, **47**:485 (1951).
208. Rushton, J. H.: *Ind. Eng. Chem.*, **44**:2931 (1952).
209. Rushton, J. H., E. W. Costich, and H. J. Everett: *Chem. Eng. Progr.*, **46**:395, 467 (1950).
210. Rushton, J. H., and J. Y. Oldshue: *Chem. Eng. Progr.*, **49**:161 (1953).
211. Vermeulen, T., G. M. Williams, and G. E. Langlois: *Chem. Eng. Progr.*, **51**: 85F (1955).

PACKED TOWERS (FLOW PATTERN)

212. Baker, T., T. H. Chilton, and H. C. Vernon: *Trans. AIChE*, **31**:296 (1935)
213. Bertetti, J. W.: *Trans. AIChE*, **38**:1023 (1942).
214. Burke, S. P., and W. B. Plummer: *Ind. Eng. Chem.*, **20**:1196 (1928).
215. Chilton, T. H., and A. P. Colburn: *Trans. AIChE*, **26**:178 (1931).
216. Dell, F. R., and H. R. C. Pratt: *J. Appl. Chem.*, **2**:429 (1952).
217. Elgin, J. C., and F. B. Weiss: *Ind. Eng. Chem.*, **31**: 435 (1939).

218. Kirschbaum, E.: *Z. ver. Deut. Ing.*, **75**:1212 (1931); *Chem. Fabrik*, 1931, p. 38.
219. Leva, M.: *Chem. Eng. Progr.*, **43**:549 (1947).
220. Lobo, W. F., L. Friend, F. Hashmall, and F. Zenz: *Trans. AIChE*, **41**:693 (1945).
221. Mayo, F., T. G. Hunter, and A. W. Nash: *J. Soc. Chem. Ind.*, **54**:375T (1935).
222. Piret, E. L., C. A. Mann, and T. Wall: *Ind. Eng. Chem.*, **32**:861 (1940).
223. Sakaidis, B. C., and A. I. Johnson: *Ind. Eng. Chem.*, **46**:1229 (1954).
224. Scott, A. H.: *Trans. Inst. Chem. Engrs. (London)*, **13**:211 (1935).
225. Sherwood, T. K., G. H. Shipley, and F. A. L. Holloway: *Ind. Eng. Chem.*, **30**:765 (1938).
226. Tour, R. S., and F. Lerman: *Trans. AIChE*, **35**:719 (1939).
227. Weimann, M.: *Chem. Fabrik*, 1953, p. 411.

PROCESS DEVELOPMENT

228. Baekeland, L. H.: *J. Ind. Eng. Chem.*, **8**:184 (1916).
229. Barneby, H. L.: *Trans. AIChE*, **40**:559 (1944).
230. Barneby, H. L.: *Ind. Eng. Chem.*, **37**:413 (1945).
231. Buck, C., T. Hayes, and R. R. Williams: *Trans. Inst. Chem. Engrs. (London)*, **24**:44 (1946).
232. Carlsmith, L. E., and F. B. Johnson: *Ind. Eng. Chem.*, **37**:451 (1945).
233. Cooper, H. C.: *Chem. Met. Eng.*, **32**:426 (1925).
234. Darlington, C. H.: *Trans. AIChE*, **31**:506 (1935).
235. Johnstone, R. Edgeworth: *Ind. Chemist*, **26**:339 (1950).
236. Little, A. D.: *J. Soc. Chem. Ind.*, **48**:202T (1929).
237. Payne, J. W.: *Ind. Eng. Chem.*, **45**:1621 (1953).
238. Pierce, D. E.: *Trans. AIChE*, **29**:100 (1933).
239. Redman, L. V.: *Ind. Eng. Chem.*, **20**:1242 (1928).
240. Stine, C. M. A.: *Ind. Eng. Chem.*, **24**:191 (1932).
241. Villebrandt, F. C.: *Trans. AIChE*, **31**:494 (1955).
242. Wagner, C. R.: *Oil & Gas J.*, **42**:(22):73 (October, 1943).
243. West, J. H.: *Chem. Age (London)*, **28**:380 (1933).
244. Whiting, J.: *8th Intern. Congr. Appl. Chem.*, New York, 1912; *Chem. Trade J.*, **51**:322 (1912).
245. Wolford, E. Y.: *Ind. Eng. Chem., News Ed.*, **3**(22):1 (Nov. 20, 1925).

SIMILARITY AND DIMENSIONAL ANALYSIS

246. Bridgman, P. W.: "Dimensional Analysis," Yale University Press, New Haven, Conn., 1931.
247. Buckingham, E.: *Phys. Rev.*, **4**(4):345 (1914).
248. Buckingham, E.: *Nature*, **96**:396 (1915).
249. Duncan, W. J.: "Physical Similarity and Dimensional Analysis," Edward Arnold & Co., London, 1953.
250. Huntley, H. E.: "Dimensional Analysis," MacDonald & Co. (Publishers), Ltd., London, 1952.
250a. Johnstone, R. Edgeworth: *Ind. Chemist*, **6**:482 (1930).
251. Langhaar, H. L.: "Dimensional Analysis and Theory of Models," John Wiley & Sons, Inc., New York, 1951.

252. Matz, W.: "Änwendung des Ahnlichkeits-grundsatzes in der Verfahrenstechnik," Springer-Verlag OHG, Berlin, 1954.
253. Newton, Isaac: "*Principia*," Motte's translation revised by Cajori, book II, proposition XXXII, Cambridge University Press, New York, 1934.
254. Rayleigh, Lord: *Proc. Roy. Soc. (London)*, **66**:68 (1899–1900).
255. Rayleigh, Lord: *Nature*, **95**:66 (1915).
256. Riabouchinsky, D.: *Nature*, **95**: 591 (1915).
257. Rushton, J. H.: *Chem. Eng. Progr.*, **48**:33, 95 (1952).
258. Thring, M. W.: *Trans. Inst. Chem. Engrs. (London)*, **26**:91 (1948).
259. Tolman, R. C.: *Phys. Rev.*, **3**(4):244 (1914).
260. Van Driest, E. R.: *J. Appl. Mechanics*, **13**(1):A-34 (1946).
261. Williams, W.: *Phil. Mag.*, **34**:234 (1892).

STRESSES IN SOLID MEMBERS

262. Coker, E. G.: *Proc. Inst. Mech. Engrs. (London)*, 1926, p. 897.
263. Griffith, A. A., and G. I. Taylor: *Adv. Comm. Aeronautics Tech. Rept. (Brit.)*.
264. Griffith, A. A., and G. I. Taylor: *Adv. Comm. Aeronautics Tech. Rept. (Brit.)* 333(**3**), p. 920, 1917–1918.
265. Griffith, A. A., and G. I. Taylor: *Adv. Comm. Aeronautics Tech. Rept. (Brit.)* 334(**3**), p. 910, 1917–1918.
266. Griffith, A. A., and G. I. Taylor: *Adv. Comm. Aeronautics Tech. Rept. (Brit.)* 392(**3**), p. 938, 1917–1918.
267. Griffith, A. A., and G. I. Taylor: *Adv. Comm. Aeronautics Tech. Rept. (Brit.)* 399(**3**), p. 950, 1917–1918.
268. Nádai, A., and A. M. Wahl: "Plasticity," pp. 130, 138, McGraw-Hill Book Company, Inc., New York, 1931.
269. Neubauer, T. P., and O. W. Boston: *Trans. ASME*, **69**:897 (1947).
270. Prandtl, L.: *Z. angew. Math. u. Mech.*, **3**:442 (1923).

INDEX

Page references in **boldface** type indicate main discussions